Precision Pendulum Clocks
The Quest for Accurate Timekeeping

Derek Roberts

Major Contributors
**Jonathan Betts, John Martin,
Alexander Stewart, Denys Vaughan**

Schiffer Publishing
4880 Lower Valley Road, Atglen,

Cataloging-in-Publication Data

Roberts, Derek.
　　Precision pendulum clocks: the quest for accurate timekeeping in England/by Derek Roberts.
　　　p. cm.
　　　ISBN 0-7643-1636-2
1. Clock and watch making—England—History. 2. Clock and watch making—France—History. 3. Clock and watch making—Germany—History.
I. Title.
TS543.E54R63 2003
681.1'13'0942—dc21

2002156717

Copyright © 2003 by Derek Roberts

　　All rights reserved. No part of this work may be reproduced or used in any form or by any means—graphic, electronic, or mechanical, including photocopying or information storage and retrieval systems—without written permission from the publisher.
　　The scanning, uploading and distribution of this book or any part thereof via the Internet or via any other means without the permission of the publisher is illegal and punishable by law. Please purchase only authorized editions and do not participate in or encourage the electronic piracy of copyrighted materials.
　　"Schiffer," "Schiffer Publishing Ltd. & Design," and the "Design of pen and ink well" are registered trademarks of Schiffer Publishing Ltd.

Designed by John P. Cheek
Cover Design by Bruce M. Waters
Type set in Bodoni BT/Korinna BT

ISBN: 0-7643-1636-2
Printed in China

Published by Schiffer Publishing Ltd.
4880 Lower Valley Road
Atglen, PA 19310
Phone: (610) 593-1777; Fax: (610) 593-2002
E-mail: Info@schifferbooks.com
Please visit our web site catalog at
www.schifferbooks.com
We are always looking for people to write books on new and related subjects. If you have an idea for a book, please contact us at the above address.

This book may be purchased from the publisher.
Include $3.95 for shipping.
Please try your bookstore first.
You may write for a free catalog.

In Europe, Schiffer books are distributed by
Bushwood Books
6 Marksbury Avenue
Kew Gardens
Surrey TW9 4JF England
Phone: 44 (0) 20 8392 8585
Fax: 44 (0) 20 8392 9876
E-mail: Bushwd@aol.com
Free postage in the UK. Europe: air mail at cost.

Contents

Major Contributors 4
Acknowledgments 5
Photo Credits 6
Foreword 7
Introduction 8
Chapter 1: Finding and Keeping Time: From Stars to Pendulum By J. Betts 9
Chapter 2: The Birth of Accurate Timekeeping: The Royal Observatory By D. Roberts 14
Chapter 3: Solar, Mean Solar and the Equation of Time, Sidereal, Local
    and Greenwich Mean Time By D. Roberts 24
Chapter 4: Factors Affecting the Isochronicity of a Pendulum By D. Roberts 51
Chapter 5: Compensated Pendulums By D. Roberts 67
Chapter 6: Escapements By J. Martin 111
Chapter 7: Electric Clocks By Denys Vaughan 140
Chapter 8: Pendulum Clock Precision 1750-1960 By A. D. Stewart 163
Chapter 9: Thomas Tompion and George Graham By D. Roberts 168
Chapter 10: John Harrison By D. Roberts 190
Glossary of Terms 217
Index 223

Major Contributors

Jonathan Betts studied at the BHI's course in technical Horology at Hackney College, London, between 1972 and 1974 after which he spent a short while in a London firm restoring antique clocks and musical boxes. He then spent 5 years in business on his own account as a freelance antique clock restorer and in 1979 was appointed Senior Conservation Officer (Horology) at the National Maritime Museum. He has been Curator of Horology since 1990.

In his own time, he has been Horological Adviser to the National Trust of Great Britain since 1980; Horological Adviser to the Wallace Collection (London) since 1983; Curatorial adviser to the Harris Collection at Belmont in Kent since 1984 and Curatorial Adviser to the Worshipful Company of Clockmakers since 1992. He is a member of the Clocks Committee at the Council for the Care of Churches.

He has written numerous articles, letters and reviews in the Horological and antiquarian press; has, over the years, given many lectures on Horological subjects, including at Harvard University and the Royal society in London and makes regular radio and television appearances on the subject of Horology. His present work at the museum includes a catalogue of the collection of marine chronometers and precision watches and he is writing a biography of the great horologist and polymath Lt. Cdr. R. T. Gould (1890-1948). He is a Huntington Fellow at the Mariners Museum at Newport News Virginia, and has recently catalogued their collection of chronometers.

John Martin was born in 1931, educated at Sevenoaks School in Kent from where, in 1948, he went to Vickers-Armstrongs in Crayford as a Pupil Engineer Apprentice.

Following graduation and national service, some fifteen years was spent as a draughtsman and in heavy engineering sales. A deep interest in horology then developed and, in 1973, he set up in business restoring antique clocks with a specialty in precision pendulum clocks. This was soon followed by the formation of a partnership business with Derek Roberts to design, build and sell a range of regulator clocks in the name of Martin and Roberts, Clockmakers.

Over the years he has acquired a wide practical knowledge of mechanical escapements, has written and lectured on the subject of precision pendulum clock design and is a Liveryman of the Worshipful Company of Clockmakers.

Derek Roberts has been professionally involved in the world of horology for some thirty-five years, running a world-renowned business in Tonbridge, thirty miles south of London. He gained his Freedom of the Worshipful Company of Clockmakers in 1977 and was elected to the livery in the following year. He and his highly skilled team of clock and cabinet makers buy, research, restore and sell a wide range of clocks, all of which are carefully documented and photographed. It is these records that have formed the basis for his books that include *Carriage and Other Travelling Clocks*, *The British Longcase Clock*, *Mystery, Novelty & Fantasy Clocks*, *British Skeleton Clocks* and *Continental & American Skeleton Clocks*.

He has held several major exhibitions, the most important of which was 'Precision Pendulum Clocks' in 1986. It was the one hundred twenty-five page catalogue which accompanied this exhibition, that formed the foundation on which this book was built, although his interest in this aspect of horology goes back much further.

Alexander D. Stewart, born in Glasgow in 1933, graduated in geology and physics from the University of Liverpool 1956, followed by a PhD in 1960. He was elected a Fellow of the Geological Society of London in 1964. He did research in the eastern Alps, based in the Department of Petrography, University of Munich, 1960-1963. This was followed by research in palaeomagnetism and stratigraphy in the Geology Department, University of Reading 1963-1968. He was appointed to teach tectonics in this Department in 1968. He retired early in 1987 to live in Italy, and spent the following fifteen years doing research in sedimentary geochemistry. His most recent publication (2002) is a Geological Society memoir on the Torridonian rocks of Scotland. He was fascinated by the history of clockwork as a boy, but it was only in 1992 that he bought his first regulator clock. Since then, he has published several papers on the design and precision of this and other pendulum clocks.

Denys Vaughan is a physicist who was formerly senior curator in charge of the Time Measurement collection at the Science Museum in London. He was responsible for mounting the exhibition "Electrifying Time", which was held at the Museum in 1977 to mark the centenary of the death of Alexander Bain. Following his retirement he has maintained his association with the Museum as an honorary Research Fellow in the Science Museum Library. He is a member of the Antiquarian Horological Society and edited its journal, *Antiquarian Horology* for six years during the 1990's. He has published articles on a wide range of metrological subjects and is currently working on a revised edition of Hope-Jones' classic work, *Electrical Timekeeping*.

Acknowledgments

This book could not have been written without the co-operation and support of numerous different people and organisations involved in the field of horology. Some have assisted by supplying information, photographs, articles and access to and details of various regulators scattered over many countries, whilst others have contributed chapters on the areas of which they have specialised knowledge. John Martin for instance, who has made the movements for some forty regulators, produced by the partnership of Martin & Roberts over the last twenty five years, has studied their design in depth and provided an excellent account of the development of the escapements used from the invention of the pendulum in the 1650s through to the first part of the 20th century.

Denys Vaughan, ex Science Museum, has brought his profound knowledge of electrical clocks to bear by describing the contribution they made in the field of precision pendulum clocks. Similarly, Sandy Stewart, who has spent many years studying their accuracy, has provided an excellent summary on this subject.

Jonathan Betts, Curator of horology at the Royal Observatory, National Maritime Museum, has provided an excellent introduction to the book by writing on the various ways of recording and measuring time and the practical use to which that information has been put, and indeed proved invaluable, over the centuries. He has also contributed a very well researched account of Earnshaw's regulators.

Don Saff has produced a fine chapter on America's contribution to accurate timekeeping, a subject on which all too many of us on this side of the Atlantic have a relatively limited knowledge, and George Feinstein has allowed us to reproduce his article, in a somewhat updated and modified form, on Fedchenko's regulators, which was first published by the NAWCC.

It has been left to Philip Woodward to conclude the book by summarising the continuing development of the precision pendulum clock in the last 50 years, including the most recent advances. He also very kindly edited Professor Hall's notes on his regulator, known as the Littlemore Clock.

One of the most generous gestures was that of Hans Staeger, whom I have known and respected for many years. He published his excellent book *100 Years of Precision Timekeepers from John Arnold to Arnold & Frodsham 1763-1862* in 1997. Not only did he give me free access to and permission to use any of the material in that book which might be of assistance to me; he also scoured all his records and produced much other helpful information and photographs.

Dieter Riefler was equally generous in allowing me to reproduce anything in his book on Sigmund Riefler which might be of interest to the readers, and Hans-Jochen Kummer and Herbert Dittrich were just as helpful so far as their book on Ludwig Strasser and the precision pendulum clocks which Strasser & Rohde produced were concerned.

Throughout the writing of the book I have contacted Christian Pfiffer-Belli on numerous occasions for assistance and invariably it was forthcoming with the minimum of delay and fuss; nothing ever seeming to be too much trouble.

To try and reduce any mistakes to a minimum Christopher Hurrion and John Martin have both read through the entire text for me, a long and arduous task which I have much appreciated and Jonathan Betts has passed helpful comments on it, and others, such as Dieter Riefler, have checked specific chapters.

John Martin produced all the line drawings which appear in his chapter and also others which are used elsewhere and further line drawings were supplied by Robin Edmunds.

So far as practical all new photography has been produced for the book, much of it by the author. Those who have given me access to their collections include the Trustees at Belmont, where I photographed all the precision pendulum clocks in the Lord Harris' collection; The National Maritime Museum, where I took pictures of the 1713 and 1728 Harrison's in the collection of the Worshipful Company of Clockmakers and also the Royal Astronomical Societies' Harrison.

Lord Yarborough generously allowed me to take detailed photographs of Harrison's clock at Brocklesby Park and Professor Hall's wife and son permitted me to take pictures of the Littlemore clock. My sincere thanks also go to the private collectors, etc. who gave me access to their treasured possessions but for reasons of security cannot be mentioned.

Jeremy Evans of the British Museum kindly arranged to have several of the regulators in their collection re-photographed and indeed supervised the taking of these to try and make certain they illustrated all the salient points. Jonathan Betts of the National Maritime Museum was equally helpful in this respect, both with regard to the clocks at the National Maritime Museum and also those at Belmont.

As with nearly every other book I have written, Michael Turner of Sotheby's very kindly gave me free access to their Photographic Library, ably assisted by Alex Barter. Stefan Müser of Auktionen Dr. H. Crott was also most generous in searching through his records and supplying many photographs, particularly of German precision pendulum clocks and Antiquorum in Geneva were equally helpful in permitting me to use illustrations of any of the precision pendulum clocks I required which had passed through their hands since their inception. Sadly the two double pendulum Janvier's came up for sale just too late to be included in the book.

Many Museums, both in this country and overseas, kindly assisted; for instance the National Museum of Scotland retrieved clocks from storage so that they could be photographed for the book. Prescott Museum supplied me with an excellent set of photographs of regulators by Condliff and Litherland; Musée International D'Horlogerie, La Chaux-de-Fonds, gave me photographs of their Le Roy tank regulator and other helpful pictures, the Science Museum let me have photos of their longcase by Harrison; Cumming's barometer clock and the Vulliamy in their care, whilst Royal Collection Enterprises supplied me with excellent photos of the fine regulators in the Queen's Collection. The pictures supplied by the Patrimonio Nacional, Madrid were particularly helpful in that they displayed clocks by English makers such as Shelton, the like of which were never seen on the home market.

The co-operation of the Worshipful Company of Clockmakers in supplying pictures of many of the famous makers of precision pendulum clocks and also their permission to photograph their two clocks by Harrison added much to the book as did the permission of the British Horological Institute to reproduce various pictures held by them including the premises of several clockmakers. Similarly the permission of the Royal Astronomical Society to photograph their Harrison, probably the last one he made and certainly the finest, was an important contribution.

This book could not have been written without the terrific effort put into the production of the manuscript by Rosemary Freeman and the overcoming of the numerous queries and problems which have arisen. She was very ably supported in this by Elizabeth Stracey. I have also received much personal support from Duncan Greig and Robert Wren.

Lastly, but by no means least, sincere thanks are due to Peter and Nancy Schiffer for giving me the opportunity to write this book and then seeing it through to such a successful conclusion.

The tolerance of my wife these last few years in seeing me disappear into my study at 8.00 am and often not emerging until 8 or 9 at night, except for sustenance, is almost beyond comprehension, as was her constant support for this venture, without which it could never have been written.

Photo Credits

Arnfield, J. Figs. 5-25B.
© The Bridgeman Art Library/Worshipful Company of Clockmakers. Figs. 4-1, 5-8, 10-10.
© Copyright The British Museum. Figs. 2-4D, 3-7, 3-16A-D, 5-1, 5-2A-C, 5-43A-C, 9-3A-C, 9-4A-C, 9-15A-C.
© Christie's Images Ltd (2002). Figs. 5-15A,B, 5-16A,B, 5-19, 9-10.
Crott, Dr. H. (Stefan Muser). Fig. 7-20B, 7-21.
Deal Time-ball Museum. Figs. 7-14.
Earl of Yarborough. Figs. 10-4, 10-5A-K.
Fitzwilliam Museum, Cambridge. Figs. 9-5A,B, 9-6.
Ford, Jack. Fig. 7-14.
Harris (Belmont) Charity. Fig. 6-16.
Lee, A. Fig. 9-12.
Mallinder, David. Figs. 3-17A-C.
Martin, John. Figs. 4-10, 6-3, 6-4, 6-5A,B, 6-6A,C,D, 6-7A,C, 6-8, 6-9, 6-10, 6-11, 6-12, 6-13, 6-14A-C, 6-15A,B, 6-17A, 6-18A, 6-19, 6-20, 6-21A-C, 6-22A,B, 6-23A,B, 6-24.
Mathys. R. Figs. 5-46, 5-47.
Pfeiffer-Belli, C. Fig. 7-5.
Photos Musée international d'horlogerie, La Chaux-de-Fonds, Suisse Figs. 7-4A.
Museum of History of Science, Oxford. Figs. 7-16A,B.
© National Maritime Museum, London. Figs. 2-2, 2-3, 2-4A-C, 2-5, 2-7, 2-8, 7-12A, 9-13A-B.
© Patrimonio Nacional, Madrid. Figs. 9-14.
Riefler, Dieter. Figs. 5-6, 5-36, 5-37, 5-39, 7-9A,B.
Roberts, Derek. Figs.3-2A,B, 3-3, 3-4, 3-5A,B, 3-6, 3-9A-C, 3-10A,B, 3-11A,B, 3-12, 3-13A-C, 3-14A,B, 3- 21A-C, 4-4, 4-8, 4-9, 4-12A,B, 4-13A-C, 4-14A,B, 5-3B, 5-4B, 5-5, 5-9, 5-10, 5-22, 5-23, 5-26B, 5-28A,B, 5-30A,B,C, 5-31, 5-32A,B, 5-33, 5-35, 5-38, 5-44, 5-45, 6-1, 6-2A,B,C, 6-6B, 6-7B, 6-18B, 9-1, 9-8A-D, 9-9, 9-11A-C, 10-1.
Roberts, Derek. (Photographs taken by him) Figs. 6-16 (with permission of the Harris (Belmont) Charity), 10-2A-C (with permission of Worshipful Company of Clockmakers), 10-4, 10-5A-K (with permission of The Earl of Yarborough), 10-7A-G, 10-8, 10-11A-E (with permission of Worshipful Company of Clockmakers).
The Royal Collection © 2001, Her Majesty Queen Elizabeth II. Figs. 3-8A,B, 9-7A,B.
By permission of the President and Council of the Royal Society. Figs. 2-6, 4-6.
Rudolf of Triel-Sur-Seine. Fig. 7-5.
Science & Society Picture Library. Figs. 6-17B, 10-3A-C, 10-9.
Sotheby's. Figs. 9-2A-C.
Staeger, Hans. Figs. 3-15A-E, 5-11, 5-12, 5-13, 5-14, 5-21.
Time Museum. Fig. 10-6.
Worshipful Company of Clockmakers. Figs. 4-1, 5-8, 10-2A-C, 10-7A-G, 10-8, 10-10, 10-11A-E.

Whilst every effort has been made to acknowledge each photograph correctly difficulties have inevitably occurred; for instance because pictures have reached us unlabeled or sometimes second or even third hand which may well be 40-50 years old. Should we have credited any illustration incorrectly then our sincere apologies are extended to the owner of that picture.

Foreword

It is with great pleasure that I agreed to write a Foreword to Derek Roberts' latest – and largest – book; or should it be books?

I have known Derek for twenty-five years, since I first walked into his shop to acquire a modest clock. He has only moved next door in all that time but during the many years he has been interested in clocks, from long before he started to buy and sell them, he has recorded them, both with written descriptions and photographs. Even when the clock Derek has bought and sold has long since gone, the records and photographs he always takes of anything interesting that crosses his path remain, and it is on these records that he draws as a starting point for his books.

We have seen the results of all this research in his previous books on various aspects of timekeeping. The present book, however, takes us to quite different levels. Anyone who knows Derek well will be aware that, for him, the most interesting and rewarding clocks are those concerned with precision timekeeping. I can remember seeing clocks from a Tompion equation clock down to a Riefler clock in his shop, and very nearly every important precision clockmaker in between. Over many years, his annual exhibitions have concentrated on precision timekeeping more than on any other aspect of horology. We have finally reached the heart of Derek's enthusiasm and this is apparent from the detail within this book.

There is a change of format, this time, however. Previously Derek has written all the chapters himself. For this book, although most of the chapters are "his own work," he has decided to approach other eminent horologists for their specialized contributions on various aspects of precision timekeeping in order to assemble the best scholarship he can, into what will be a classic work of reference on the subject. There have been other books on precision timekeeping before, none, of course matching Lt. Cdr. Rupert Gould's ground-breaking *The Marine Chronometer* of 1923 but this present work must rank as one of the most comprehensive ever assembled on the subject.

As always, the quality of the photographs, each worth dozens of words of text, is excellent. Derek has his own studio from which to work, with equipment of the highest caliber but to capture the essential elements of a clock or watch needs more than the skills of a mere cameraman and these photographs speak for themselves.

Horologists from John Harrison onwards (if not before him) seem to have an inexorable wish to obfuscate and complicate to the confusion rather than the enlightenment of the reader. Derek succeeds in taking the reader through the convoluted paths of precision timekeeping in a comprehensible way.

This book may represent the finality of Derek's endeavors in the world of horological literature. No one interested in precision horology (what other sort is there?) will want to be without this book.

Christopher Hurrion
Master, The Worshipful Company of Clockmakers 2003

Introduction

Man's quest for accurate time-keeping probably began when he first started recording the height of the sun in the heavens or the length of the shadow cast by a tree or a mountain, and this in due course led to the shadow clocks such as Cleopatra's Needle which now resides on the Thames Embankment and a myriad of other timepieces based on the same principle, of which the garden sundial is probably the best known.

Astronomers were in the forefront of recording time, well before the birth of Christ, so that they could note the passage of various heavenly bodies and then calculate their relative positions and when they might reappear, and this close link between astronomers and accurate timekeeping is as strong today as it was many thousands of years ago.

Horology, and in particular the birth of public clocks, took a big leap forward with the invention of the first mechanical clock towards the end of the 13th century, the most important component of which was the "escapement," a device which releases a train of wheels at set intervals, which are used to indicate time. These employed a verge escapement and a foliot or balance wheel.

The next major development was the use of the swing of the pendulum, initially just to give time intervals, such as a metronome still does today, but later to be incorporated in a clock. This occurred c.1657 and within another 12-13 years a further leap forward in the quest for accurate timekeeping had been made with the birth of the anchor escapement and the seconds beating pendulum which accompanied it, and this is where our story starts.

These inventions were adopted more rigorously in London than anywhere else, in large measure because of the arrival on the scene at that time of many brilliant men, of whom Newton, Hooke and Wren are but three examples. They were joined by clockmakers such as Fromanteel, Tompion, Knibb and Clement who assisted them by making instruments and providing timekeepers. Probably the most remarkable achievement in this area was the construction of Tompion's year clocks for the new observatory at Greenwich that enabled Flamsteed to prove that the earth rotated at a constant rate.

It was not to be long before the leading clockmakers were aware that although they had made an enormous leap forward in timekeeping there were further problems to be overcome and indeed it was to be more than 200 years before these were largely resolved. Foremost amongst them was the variation in the length of a pendulum with fluctuations in temperature; the second was the effect of changes in barometric pressure and the final problem was to devise a way of always impulsing a pendulum by exactly the same amount so as to maintain a constant arc of swing.

It is the story of how the leading clockmakers, primarily of England, France and Germany, struggled to overcome these problems over the best part of three centuries, in the end achieving an accuracy of around 0.00001 of a second a day, which we have tried to tell here.

Chapter 1
Finding and Keeping Time
From Stars to Pendulum
By Jonathan Betts

For many years before accurate clocks were a reality, theoreticians were aware of the possibilities for improving Astronomy, should such a machine be created and it is not surprising that this is the area where accurate timekeeping first came into its own. The plethora of books recently published on the subject of Longitude all remind us that as early as 1514 Johann Werner of Nuremburg postulated the "lunar-distance" method of finding one's longitude at sea, and early in the seventeenth century a method employing observations of the moons of Jupiter (using the newly invented telescope), was also being researched. A major stumbling block to the former method however was that it presupposed one had accurate charts of the stars and the motions of the moon, and these could only be made if an accurate timekeeper were available. Similarly, observations of Jupiter's moons needed an accurate timekeeper as well as a telescope, and the very best one could expect from contemporary clocks controlled by verge escapement and foliot, or balance wheel, was about +/- 15 minutes a day. One early attempt at improving such timekeepers was the invention, by the Swiss-born clockmaker Jost Burgi, of the cross beat escapement in the late 16th century. Clocks by Burgi with this escapement and with a minute, and even a seconds hand, were tried at the observatory at Kassel and, in 1586, apparently enabled the positions of a number of stars to be established with considerably greater accuracy than had been managed before.[1] How much the cross beat actually improved the timekeeping is questionable though, given that the oscillator(s), like the earlier foliot and balance, had no natural frequency of their own. The great Danish astronomer, Tycho Brahe, using two clocks by Burgi in his observatory in Uraniborg in the 1590s, found they were too inaccurate to be useful.

The Pendulum

The pendulum of course does have a natural frequency, dependent on its effective length, and it is in this invention that the roots of the "precision clock" or *regulator* are to be found. Although it is said that the ancient astronomers of the East counted the beats of a freely swinging pendulum to time astronomical events,[2] the same practice, employed by the great Italian astronomer Galileo Galilei can be considered as the first use of a pendulum as a timekeeper in "modern times." Again, it was in attempting to develop a means of finding longitude at sea, that Galileo primarily made the observations, his intention being to produce accurate tables for the motions of the moons of Jupiter. It was then a logical step for Galileo to propose, in 1637, that a geared mechanism might be used to record the swings of the pendulum, and a further logical step for him to realize that the geared mechanism, if driven with a weight, might also give pushes to the pendulum, an idea he communicated to his son in 1641, one year before his death. But history relates that it was the Dutch scientist Christiaan Huygens, working with the Hague clockmaker Salomon Coster, who, in early 1657, first created a practical, pendulum-controlled clock. In conceptual terms then, Galileo applied clockwork to a pendulum, while Huygens applied the pendulum to clockwork and it was the latter which resulted in the practical design, the true ancestor of the regulator.

The introduction of the pendulum to clockwork coincided with the beginning of that extraordinary intellectual flowering which became known as Europe's scientific "Golden Age". Thus, as the first instruments truly capable of accurate time determination made their appearance, so did the first "modern" astronomical observatories, still primarily with the remit to research and develop positional astronomy as a means of improving cartography and finding longitude at sea. Those at Paris, founded in 1667 and Greenwich founded in 1675, were at the forefront of this work, though it was the latter which took the lead in creating the world's first accurate star charts and commissioning "state of the art" regulators to serve that work. So, in considering how the mechanical regulator was developed and used in observatories over the following two centuries, the example of Greenwich is probably the most representative.

Positional Astronomy

The theory of drawing up accurate charts of the heavens could not be simpler in principle, it was basically the same as making a chart of the globe. For every celestial body, the position, equivalent to its latitude and longitude, would be established by noting the two coordinates as the body reached its highest point in the sky. This would naturally occur on one's *meridian*, an imaginary line across the heavens stretching from the southern horizon up, over the observer, to the northern horizon. The two coordinates are height above the horizon, called the *Declination* (like latitude and recorded in degrees, minutes and seconds of arc) and, because the earth is revolving, the time at which the body crosses the meridian, which is called the *Right Ascension* or *RA* (similar to Longitude and normally recorded in hours, minutes and seconds of time). The practice of making such observations involves the use of a telescope, fixed in a North/South plane so that it can only scan up and down the meridian. Telescopes like this became known as a *Transit Instrument* and this, and the regulator, sometimes known as the *Transit Clock*, and usually rated to sidereal (star) time, were the two essential "laboratory tools" of the astronomer.

There was one assumption with this procedure however, which was that the earth naturally spins on its axis at a uniform rate. If this was not so, and the rate was sometimes a little faster, sometimes a little slower, then recording the RA of a celestial body would be meaningless. Proving this assumption was the preliminary task for John Flamsteed, the first Astronomer Royal at Greenwich; and in just two years this was done. Using the two year-going clocks by Tompion, with long pendulums and dead beat escapements, mounted in the panelling in the Great Room (known as the Octagon Room today, Fig. 2-7) and which were regularly compared with transits of the star Sirius, Flamsteed was able to prove the Earth's "Isochronicity", as he called it. Of course, today we know this not to be *exactly* so, but Flamsteed was satisfied that continuing astronomical observations would be meaningful.

Another phenomenon which urgently required better tabulation was the *Equation of Time*. This is the difference between *Mean Solar Time* ("clock time") and *Apparent Solar Time* ("sundial time" or time derived from the sun crossing ones meridian). Owing to the inclination of the Earth's axis, and its elliptical path around the Sun, during the course of the year mean and apparent time differ from each other from day to day. Fortunately, every year they will do so by almost exactly the same amount on any given day, so a table can be drawn up to enable clocks to be set using a sundial throughout the year. Thus, one of the first roles for the new, accurate pendulum clocks would be to enable other clocks of the same kind to be set accurately when in use. Using pendulum clocks, Christiaan Huygens had determined and published figures for the Equation of Time in 1665 and Flamsteed himself had produced tables in manuscript form the following year (then published in 1673), but Flamsteed needed to confirm these data once he had established the Earth's "isochronicity," and this was another role the Tompion clocks successfully played.

Once the foundations for the research were thus laid the really hard work of observing began, a task which was to continue far longer than anyone had imagined. In the 1690s alone Flamsteed recorded that, in addition to a number of other astronomical duties, they had made 25,000 observations and established the positions of about 4,500 stars.[3] And it was hard work; day (observations of Sun and Moon and a few brighter stars) and night, recording transit after transit, without lighting at night and only a little heating being allowed in the Winter which would otherwise disturb the clarity of the atmosphere.

In fact the "requisite tables" necessary for determining the longitude were to take many more years. John Harrison tells us that in about 1728-30, when he visited Flamsteed's successor, Edmund Halley (of comet fame), he was confidentially informed by George Graham that Halley was beginning to despair of ever completing the tables. Halley, who was *"then becoming quite tired of it, or thoroughly satisfied as touching the impossibility of its ever doing any certain good..."* therefore *"received me the better."*

After Flamsteed's death in 1719, all his clocks (mostly by Tompion), including the two year-going examples in the Great Room and the "angle clock" (Fig. 2-8), used for a short while by Flamsteed for recording RA in terms of angle rather than hours minutes and seconds of time, were removed and sold. Halley was provided with funds by the Royal Society to re-equip the Observatory and it was in the 1720s that the first three regulators by Graham were purchased, one 8 day clock which was Halley's principle transit clock, and two month going ones, all originally with simple, uncompensated pendulums. The most famous Graham clock at the Observatory was however the one bought by the next Astronomer Royal James Bradley in 1750. As a month going clock, it was the third of this type and was then referred to as "Graham No. 3." Graham 3 (Fig. 9-13) was to be the principal transit clock during the reigns of Bradley (1742-1762), his successor Nathanial Bliss (1762-1764), then Nevil Maskelyne (1765-1811) and the first ten years with John Pond (1811-1835).

The work at Greenwich: Taking the Time of a Transit

The Astronomers Royal had, at various times, a number of assistants, but the work was unrelenting and required a very meticulous and determined kind of mentality. In later years, an assistant of Nevil Maskelyne's wrote:

Nothing can exceed the tediousness and ennui of the life the assistant leads in this place, excluded from all society; except perhaps that of a poor mouse which may occasionally sally forth from a hole in the wall, to seek after crumbs of bread dropt by his lonely companion at his last meal... He is also frequently up three or four times in the night (an hour or two each time), and always one week in the month when the moon souths in the night time....[4]

With careful observing it was easily possible to determine the time of the transit of a star to well within a tenth of a second, and a simple explanation of how this was done may be useful:

The method used eye and ear together, the eye to observe through the telescope and the ear to record the time from the regulator; for this reason, astronomers have always preferred regulators with a good, loud tick. A minute or so before the observation takes place the astronomer would settle down in front of the transit instrument, set it to the correct declination, and note the time (hour and minutes) on the regulator. As the star (or Sun, moon or planet) entered the field of view, he would look at the regulator dial and, noting the exact minute, would start to count the seconds ticks, then transferring his eye to the eyepiece. The exact moment of transits would be recorded from the sound of the regulator's ticks. Fearing that the observer may lose count of the seconds, Nevil Maskelyne, in the 1760s, introduced the concept of the Journeyman, or "Assistant" clock, (usually thirty hour clocks showing only minutes and seconds) which sounded a bell at the top of every minute (Fig. 12-10A), giving the assistant notice to start counting. Quite a number of these clocks were made for observatories like Greenwich, but very few survive today in their original form.

The eyepiece of Maskelyne's transit telescope contained five carefully placed vertical "wires." These naturally had to be as fine as possible. A few years later (but probably not in Maskelyne's time) these were created using the thread from a *spider's web*, one of the assistant's more unusual tasks being to collect thread from spiders' webs in the astronomer's garden behind the observatory whenever new wires were needed!

The five, equally spaced wires crossed the field of view, top to bottom, and the observed star was seen to pass behind them, appearing to travel from right to left when viewing the majority of the stars (in the southern sky), as the telescope inverted the image. The time at which the star coincided with each of the five wires was carefully noted, one by one, an average of the five wires then producing a more accurate result than if the central "meridian wire" alone had been observed. Stars were seen as a bright "pin point" of light, without a perceptible disc, and would often be "...surrounded by little flashing, vibrating rays of colour...."[5] A star appeared to move relatively quickly: a full transit across the five wires would take about two and a half minutes for a star at average declination. The pin point of light and the fineness of the wire were such that within two beats of the clock (i.e. one full second) the viewer could easily observe the star pass from one side of a wire to the other; indeed he could, "in his mind's eye" even divide the space between the two ticks into ten (i.e. tenths of a second of time) and apportion those tenths either side of the wire. In this way, very accurate timing of the transits of stars was possible and, if a number of "clock stars" were observed during an evening's viewing, an average of their readings would enable the transit clock to be checked to a very high degree of precision.

When viewing larger bodies, such as the Sun (a dark filter was used over the eyepiece) or Moon, both "limbs" (edges) would be observed as they passed the wires. The Mean Solar Standard clock, a separate regulator specifically to maintain GMT in the Observatory, would also be corrected by transit of the Sun, the time recorded first on the Transit (sidereal) clock and then corrected to Mean Solar Time.

On occasions an assistant failed to do the work correctly, sometimes with unfortunate consequences. In 1796, in a sad, but revealing note in his annual report (Greenwich Observations) Nevil Maskelyne wrote:

I think it neceffary to mention that my Affiftant, Mr. David Kinnebrook, who had obferved the tranfits of the ftars and planets very well, in agreement with me, all the year 1794, and for the great part of the prefent year, began, from the beginning of Auguft laft, to fet them down half a fecond of time later than he fhould do, according to my obfervations, and in January of the fucceeding year, 1796, he increased his error to 9/10 of a second. As he had unfortunately continued a considerable time in this error before I noticed it, and did not seem to me likely ever to get over it, and return to a right method of obferving, therefore, though with reluctance, as he was a diligent and ufeful affiftant to me in other respects, I parted with him.

Maskelyne comments on the instruments used and then explains the exact method by which the star transits are timed (as interpreted above) enabling him to check the error of the transit clock, (Graham 3) the rate of which is then recorded to two decimal places:

The joint excellence of the clock, tranfit inftrument, and method of obferving, when properly attended to, have been the means of first difcovering this error; and then afcertaining the quantity of it. The clock, originally made by Graham, has been fucceffively improved by Arnold and Earnshaw, and the pendulum hung from the pier detached from the clock work: the transit inftrument made by Bird has been improved, by fubftituting an achromatic objectglafs inftead of the common one, and by putting the finest wires, of the thickness of the 1/1000 part of an inch, in the focus..... The method of obferving, introduced by Bradley, of noting the proportional diftance of the ftar from the wire at the two beats immediately preceding and following the tranfit acrofs the wire, has been carefully ufed, and thence the tranfit fet down to the tenth of a fecond. I cannot perfuade myfelf that my late Affiftant continued in the ufe of this excellent method of obferving, but rather fuppofe he fell into fome irregular and confufed method of his own; as I do not fee how he could have otherwife committed fuch grofs errors. However, this unfortunate inftance fhews the neceffity of firft duly underftanding, and then closely adhering, to this method of obferving ..... The great thing is to aim always at the truth, and avoid any partial method of obferving ..... The difficulty of attaining the defired exactnefs arifes from various caufes; fometimes from the great faintnefs of the object, fometimes from its over brighnefs, a tremor or undulation owing to a bad ftate of the air, or the quick motion of the ftar through the field of the telescope occafioned by the great magnifying power, and fometimes from flying clouds. We fhould obferve with all our attention, when the ftar comes near the wire, and fix (as if we could mark down) the apparent places of the ftar in the field of the telefcope at the two beats of the clock before mentioned, and thence eftimate and note down the proper fecond and tenth anfwering to the tranfit acrofs the wire. If we are not quick in fixing the place of the ftar at the beat, we fhall be apt to affign to it too backward a place in the telefcope and confequently to reckon the time of the tranfit too great, as my late Affiftant did. Sometimes, efpecially when the ftar was very diftinct and fteady, I have, by the obferved rate of its motion during the forgoing fecond or feconds, carried my eye on, and anticipated the place the ftar was to occupy at the fucceeding beat of the clock, and only corrected it if neceffary (which was feldom the cafe) by the actual obfervation. A good ear feems, in this kind of obfervation, to be almoft as ufeful as a good eye.

Occasionally, the thousands of dry entries in the annually published *Greenwich Observations* are brought to life by everyday remarks, sometimes so ordinary as to be almost surreal. For example, Maskelyne's assistant Thomas Firminger recorded, when taking a transit of the Sun on August 28th 1806:

A little before the last limb came to the last wire in the transit instrument the beat of the clock was all at once silent. Upon looking up immediately at the dial, I perceived the seconds hand to stand still, or nearly so; but before I could get up to the clock to see what might be the cause of its stopping, it went off again and I observed a small insect to run away from the place at which the seconds hand stood still. The stopping of the clock was therefore occasioned by the insect running across the dial, and being caught in the space between the seconds hand and dial, left for the free motion of the seconds hand. The insect, being a little larger than this space, got wedged in between the seconds hand and dial as the hand was passing round, and being but little larger than the space, yielded to the pressure of the hand, which passed over the insect and then went off again as before. I opened the case of the clock and caught the insect, which I found to be of the beetle kind, called by naturalists Cerambic. These insects are to be found very common in most kinds of wood, but particularly in old furniture, wainscotting and such wood as has been a long time in use, into which they eat small holes ... How long the insect had been in the clock case is of course uncertain. From the above mentioned circumstance the observed transit of the Sun this day is useless, on which account it has not been inserted in the book.

The clock was, of course, Graham 3, though no signs of woodworm can be found in the case today!

The Drum Chronograph

The same method of observing continued until the middle of the nineteenth century, when the formidable George Biddell Airy, 7th Astronomer Royal, introduced a new electrical system for more accurately and consistently noting transits. In fact, the idea was first developed in the U.S. in the late 1840s by the scientist John Locke in conjunction with astronomer Sears Walker for the U.S. Coast Survey and acknowledging this, Airy always referred to it as the "American Method."[6] The system, which continued in use right up until the mid twentieth century, employed what was subsequently called a "drum chronograph", a large (approximately 30cm diameter) drum about 60cm long and covered with a sheet of paper. The drum was revolved at a relatively constant rate and a stylus scanned across it over the course of several hours. The stylus, operated by an electrical coil, was caused to make a mark (in the early years the "stylus" was in fact a "pricker" which punched a tiny hole) on the paper every second by a pulse from electrical contacts fitted to a regulator with a seconds pendulum (the first one at Greenwich was the transit clock by Hardy), with the top of the minute signified by the absence of a mark. This created a long timescale of marks in an extended "spiral" (helix) round the paper, representing the observing period, with every second marked accurately upon the paper. While the chronograph was running, the stylus could also be activated by a contact mounted on the eyepiece of the telescope: the moment of transit could be sent directly to the chronograph and appeared on the paper along with the timescale. A whole evening's observing would thus be recorded on the chronograph without needing to hear, or even see the regulator. It was only when observing was finished (or the following day), that the assistant would remove the sheet from the drum and note down the exact times of the observations, reading off to the nearest second and then carefully measuring between the two seconds marks to record the exact moment the observation was made. The system removed much of the risk of human error in determining the actual time of a transit, but the split second at which the transit was recorded was still subject to slight variation from one observer to another, depending on their rate of reaction on the electrical key. This was a phenomenon recognized by Airy who referred to it as the observers "personal equation", and it was something he did his utmost to reduce among his Assistants.

The system was such an immediate success it was adopted by just about every professional observatory worldwide, and it would not be an exaggeration to say that the majority of transit clocks in use during the second half of the 19th century were converted to include seconds contacts, for use with chronographs of this kind. Indeed, the presence of electrical contacts in a regulator should today be seen as a probable indication of a distinguished career though, alas, many dealers and collectors still regard them as an unfortunate later alteration, the evidence of which needs removal and disguising as thoroughly as possible.

It was, of course, electricity which enabled ultimate improvements in the accuracy of pendulum clocks, making it possible to seal a free pendulum in a chamber yet ensure its swings were maintained, and it was electricity which then provided the means for superseding the pendulum as our principle time measuring device. Indeed, it was only with the introduction of quartz technology that we finally realized that, all along, Flamsteed had not been quite correct in his determination about the earth's "isochronicity."

Achievement

This seemingly esoteric story of the high accuracy pendulum clock is all very well, but one might reasonably ask "what was it actually for? How did it benefit mankind?"

There are many answers to the question and some of the benefits were much more straightforward than others. First and foremost was the original need to explore our world: to make maps and charts of its continents, coastlines and oceans and to navigate those oceans safely and efficiently. Simply put, accurate clocks enabled accurate positional astronomy which enabled accurate charts and even (using the lunar distance method of longitude timekeeping), accurate navigation itself. Nations such as Britain and France who carried out this primary research were empowered by it and simply could not have flourished politically, economically and militarily in the way they did without it. But the primary astronomical data went on to create the foundation for all subsequent astronomical researches, facilitating our understanding of our Solar System and the Universe itself. The NASA space programs would have been impossible without the body of astronomical data to support them, the roots of which lay in the eighteenth and nineteenth century observatories of Europe with their transit telescopes and regulators.

Of course, there were "lateral" benefits too. Research into accurate timekeeping technology directly improved our everyday lives. In the case of Greenwich, watchmakers had long used the Observatory as a source of checking their own clocks, and once accurate civil time dissemination began in the mid 19th century, the regulator's influence spread beyond the walls of the observatory, providing accurate time for the nation, first using the electric telegraph, then by radio and telephone. Today, few people could manage their lives satisfactorily without having a regular source of accurate time to structure the day, and it would be impossible for industry and commerce to operate effectively. Whether our global society really needs the super-accurate time standards now being developed is another question, beyond the scope of this chapter. It must be said though, that it is only because of the extraordinary accuracy of the atomic time standards now available that we can, for example, use a GPS receiver to find our position on earth to within a few centimeters. With time now by far the most accurately measured of all the standards, and the one by which the others may now be defined, there seems no end to what clocks may do for us.

References

[1] King and Millburn. *Geared to the Stars*. Bristol, 1978. p.80.
[2] Edwardes, Ernest, L. *The Story of the Pendulum Clock*. Altrincham, 1977. p.18.
[3] Forbes, Eric, G. *Greenwich Observatory, Vol. 1*. London 1975. p.49.
[4] Evans, Thomas quoted in John Evan's *Juvenile Tourist*, 1810. pp. 333 - 335.
[5] Maunder. *The Royal Observatory*. 1900. p. 157.
[6] Bartky Ian R. *Selling the True Time*. Stanford, 2000. p.36.

Chapter 2
The Birth of Accurate Timekeeping and the Royal Observatory

BY DEREK ROBERTS

The term "regulator" is such a confusing one that its use was deliberately avoided when titling this book although for convenience and brevity it will be used in the text. Many authorities would insist that any clock so described must have a "regulator dial layout" or, as some would say an "Astro," or Astronomical dial, by which is meant a center sweep minute hand and separate seconds and hour rings, usually placed below 12 o'clock and above 6 o'clock, which would exclude Tompion's year clocks at Greenwich and Harrison's regulators, most Continental examples and also many late English regulators. Arguments of a similar nature apply to such features as maintaining power, dead beat escapement and also the incorporation of a precision movement within a striking or chiming clock.

A further confusion can arise from the fact that the early precision clocks were called astronomical clocks because they were used in observatories, but when clocks started to be produced in England with astronomical indications on the dial this practice was gradually discontinued to avoid confusion.

Some will object to the term "clock" being included in the title as this is considered by many to imply the inclusion of a striking mechanism; however, it is felt that this will cause little confusion as it is a word used in common parlance to describe all timekeepers other than watches, whether striking or not.

Since his origins man has tried to measure time, probably by the built-in sense of time which all the animal species possess to a greater or lesser degree. The position of the sun in the sky and the movement of various heavenly bodies would also have given him guidance as would the length of the shadows cast by various natural bodies such as mountains and trees.

As civilization gradually progressed, a more precise method of measuring time was needed to regulate the life of the communities and decree at what hour public meetings or prayers, for instance, should take place. Thus shadow clocks such as Cleopatra's Needle which now resides on the Thames Embankment, were erected in public places and, at a slightly later date, portable examples were evolved which finally resulted in the beautiful little ivory tablet sundials and ring sundials which, in the sixteenth and seventeenth century, were the equivalent of the gentleman's pocket watch. Numerous different and fascinating forms of sundial were produced over the centuries; but undoubtedly the most common was what might be referred to as the garden sundial. In the early days of the longcase clock, this was used with the aid of equation tables to set the clock to its correct time and regulate it. Other early methods used to measure time were clepsydra (water clocks), most of which were based on the speed at which water ran out of a container with a hole in the bottom of it, and fire clocks. These last ranged from a single candle marked down its length with the hours, and lamp timekeepers where the glass oil reservoir is graduated in hours, to complex clocks in which a trail of powder gradually burnt down a channel in a container, and even alarm clocks in which, when a flame reached threads and burnt through them, the balls they were supporting fell with a loud clang onto a metal tray.

It is difficult to say when precision timekeeping by means of a mechanical clock first occurred. "Precision" is a relative term and it may well be that the clockmakers of the late thirteenth century felt that the advent of the mechanical clock incorporating a crown wheel and foliot heralded the advent of accurate timekeeping. However, these clocks were difficult to regulate as the only relatively crude methods of adjustment were moving the balance weights in or out on the foliot and varying the driving weight. The only other adjustment which would affect the going rate of the clock was the depthing of the pallets. Thus the timekeeping was directly affected by the friction in the train which probably altered appreciably as the relatively crude lubricants then available dried up or dirt got into the movement. It is difficult to know how well these early timekeepers performed but it seems unlikely that on average they held a rate better than five to ten minutes a day.

Probably the ultimate development of the verge escapement was the "Cross Beat," (Fig. 6-1) all knowledge of which had been virtually lost until comparatively recently when Professor Hans von Bertele of Vienna rediscovered it.

It was invented by Jost Burgi circa 1584, a clockmaker whose ingenuity and workmanship were of the highest order. It was gradually refined by Burgi and his pupils, but despite its marked improvement in timekeeping relative to the other clocks of the period it does not appear to have come into general use, possibly in part because of the relatively precise nature of the work required.

It is, in effect, a precision escapement, each pallet being carried by its own arbor, which permits very accurate adjustment at the point of contact of the pallet with the vertically mounted escape wheel. The two arbors are geared together and to the end of each is attached a counterbalanced arm. As the clock beats, these arms cross over each other, hence the name "cross beat." It is a fascinating action to watch.

Undoubtedly, the accuracy achieved by clocks using Burgi's cross beat, such as the famous rock crystal clock now in the Kunsthistorisches Museum in Vienna, was a great improvement on those fitted with the conventional verge escapement; however, it was only in use for a comparatively short period before the advent of the pendulum which was to supersede it.

The Birth of the Pendulum

In 1581 Galileo, then seventeen, made his famous observation in a cathedral in Pisa, that no matter how wide the arc of swing of a lamp suspended from the ceiling, the time taken to go from one side to the other was always the same. This we now term isochronous.

Although Galileo made use of this observation in astronomy with manually impulsed pendulums, it was only in his old age that he thought of applying the principle to a mechanical clock. These ideas he passed on to his son, who is said to have completed a model of his father's design, but died before it was perfected. It was thus left to the Dutch scientist Christiaan Huygens to design the first practical pendulum clock which he invented at the end of 1656 and patented, in Holland, in June of the following year, using the pendulum in conjunction with the verge escapement. He applied for a patent in France but this was not granted and no such application was made in England. He also evolved the principle of the endless rope for winding clocks whereby power is maintained whilst the clock is being wound and thus no time loss occurs.

In 1657, the first clock made to Huygens' design was constructed by Salomon Coster of the Hague, when John, Ahasuerus Fromanteel's son, was working with him (Sept. 1657-May 1658), and suddenly the accuracy of timekeepers improved from maybe five to twenty minutes a day to two to three minutes a week. Interestingly, although the first pendulum clock was made in Holland, its value was not immediately appreciated in that country and it was in England that its rapid adoption and development were to be seen. This was at least in part because the problem of isochronism had been occupying the minds of many of the leading members of the scientific world, such as Dr. Hooke, who maintained close contact with the more important clock and scientific instrument makers and thus when the Fromanteels introduced the pendulum clock to England it was greeted with open arms. They advertised it first in the *Mercurita Politicus* of 27th October, 1658 and the *Commonwealth Mercury* one month later. So far as is known, only one pendulum clock by Fromanteel is known which was made in that year (now at Lyme Park in Cheshire), and this is referred to in Michael Hurst's excellent article on *The First Twelve Years of the English Pendulum Clock*[1] which is based on R.A. Lee & R.T. Gwynn's Loan Exhibition which was held at their galleries at Bruton Place in 1969 and was accompanied by a seventy-six page catalogue. For those interested in the early evolution of the pendulum clock, the article by Edwardes and Dobson[2] is also recommended.

Circular Error or Deviation

When Huygens analyzed Galileo's work on the isochronicity of the pendulum, he found that his observation was only accurate if the swing passed through a somewhat steeper path than an arc known as a "cycloidal curve." To overcome this error, which on the verge escapement with its relatively large arc of swing could be appreciable, he suggested the use of "cycloidal cheeks" on either side of the suspension of the pendulum to modify the arc of swing; and indeed most early pendulum clocks had cycloidal cheeks.

Interestingly, M.H. Robert[3] has produced a table (Fig. 2-1), which shows that the loss of time involved per day for a pendulum swinging free in an arc of 1.0° is 1.55 seconds, and that this rises to no less than twenty-four minutes when the arc is increased to 30°, thus emphasizing how important a narrow and above all, a constant arc of swing is to precision timekeeping. The effect of circular error was greatly decreased by the invention of the anchor escapement some twelve to thirteen years later, which appreciably reduced the arc of swing of the pendulum and thus the error involved. However Huygens had, at an earlier period, used long pendulums in conjunction with the verge escapement which had a smaller arc of swing. Some one hundred or more years later, the same combination was to be seen on some of the earlier Morbier clocks.

Arc of Oscillation	Loss in 24 hours
0.5°	0.43 sec.
1.0°	1.55 "
2.0°	6.60 "
3.0°	14.80 "
4.0°	26.25 "
5.0°	41.30 "
6.0°	1.05 min.
7.0°	1.15 "
8.0°	1.35 "
9.0°	2.15 "
10.0°	2.50 "
15.0°	5.60 "
20.0°	10.70 "
25.0°	17.00 "
30.0°	24.00 "

Fig. 2-1. M.H. Robert's table taken from Saunier[3] showing the increasing loss of time with a rising arc of swing.

The Anchor Escapement

There has been much debate over the last thirty years as to who invented the anchor escapement and first fitted it to a clock; an interesting meeting on the subject was held at the Science Museum on 13th November, 1970 and subsequently reported in *Antiquarian Horology*.[4]

William Clement

For a considerable time it was thought that William Clement was the first to employ the anchor escapement, this assumption being based on two publications. The first, John Smith's *Horological Disquisitions* of 1694 in which, when discussing the early development of the pendulum clock he comments *that eminent and well known Mr. William Clement had at last the good Fortune to give it the finishing Stroke, by being indeed the real contriver of that curious kind of long pendulum which is at this day so universally in use among us.* And secondly, William Derham's *The Artificial Clockmaker* where he comments *For many years this way of Mr. Zulichem (Christiaan Huygens) was the only method, viz crown-wheel Pendulums, to play between two cycloidal Cheeks, etc…, But afterward Mr. W. Clement a London clock-maker, contrived them (as Mr. Smith saith) to go with less weight, on heavier ball (if you please) and not vibrate but a small compass, which is now the universal method of the Royal Pendulums.*

The evidence of Smith and Derham is reinforced by the turret clock which Clement made for King's College, Cambridge which is now in the Science Museum. This is inscribed Gulielmus Clement Londini Fecit 1671 and employs an anchor escapement. The records of King's College show that the clock was ordered in 1670 and had been installed and paid for by Michaelmas 1671. Although doubts have been cast on the originality of the escapement on the clock there are no signs that it has ever had a verge and thus the only other possible alternative to the anchor would be the tic-tac escapement; however it is thought that this was not introduced until c. 1673.

Joseph Knibb

Whereas the foregoing facts may indicate that Clement invented the anchor escapement, they are somewhat undermined by the fact that Joseph Knibb had installed a turret clock with anchor escapement for Wadham College before Michaelmas 1670,[5] one year earlier than Clement had completed the King's College Clock. There is a complete history of the clock up until the present day.

Christopher Wren

Christopher Wren had a close interest in this clock and probably commissioned it from Joseph Knibb and tradition has it that he designed it, although this seems somewhat unlikely; however he may well have influenced its design and indeed the arms of Wren are incorporated in the right hand spandrels.

Wren, born in 1632, entered Wadham College c. 1649 and subsequently obtained a B.A. and an M.A. (11th Dec 1653). He became of Professor of Astronomy at Gresham College in 1657 and Savilian Professor of Astronomy some three years later. From this time until he resigned in 1673 Wren visited Oxford frequently to give lectures and assist in various architectural projects. It would thus seem likely that he got to know Joseph Knibb very well at this time.

Wren had a considerable interest in clocks,[6] carrying out experiments on pendulums and devising a weather clock which Hooke subsequently added to. He is also believed to have designed a water-driven clock for Hanwell Castle.

Joseph Knibb also received an order in 1669 from Oxford University to convert the clock at St. Mary's Church to a pendulum and payment for this was made in the annual accounts ending on 16th Sept 1670.[5]

A further indication that Joseph Knibb may be the originator of the anchor escapement are the longcase clocks by him with an early form of the escapement which have been discussed by Christopher Greenwood.[7] In one of those illustrated, the escape wheel is cut as a train wheel but with the back face of each tooth taken away to facilitate a clean drop. Another clock, with a similar escape wheel but a heavily restored movement, may be seen in the Gershom Parkington Museum, Bury St. Edmunds and a completely original one with grande sonnerie striking is at Belmont.

Robert Hooke

One other name should be mentioned in connection with the anchor escapement, Robert Hooke who, in the *Artificial Clockmaker*[8] is quoted as follows: *Dr. Hooke denies Mr. Clement to have invented this; (the anchor escapement) and says that it was his invention and that he caused a piece of this nature to be made, which he showed before the R. Society, soon after the Fire of London.* However the Journal Book of the Society No. 4 p.84 of 28th Oct 1669 refers to the presentation by Dr. Hooke of *A new kind of Pendulum Clock pretended to keep time more exactly than others for astronomical observations, so contrived that the string, being in this clock fourteen feet long and having a weight of three pounds hanging to it, is moved by a very small force, as that of a Pocket Watch, the string making its whole vibration not above a degree and going seventy weeks.*

There is no suggestion in this quotation of the employment of a new escapement.

Because the anchor escapement permitted the pendulum to have a much narrower arc it was possible to increase its length. The "Royal" pendulum was thus evolved a little over thirty-nine inches long and taking exactly one second to complete its swing, which made it a simple matter to provide a seconds' hand on clocks. Timekeeping had thus taken another great stride forward in accuracy from a few minutes a week to possibly a variation of no more than a few seconds in the same period. Even at this stage clockmakers were becoming increasingly aware of the effect of the clock's movement and in particular any variation in the power it transmitted, on the frequency of vibration of the pendulum and thus accurate timekeeping. Two factors predominate here, both of which are greatly effected by the mechanical condition of the clock's components: (a) the power transmitted to the pendulum via the escape wheel, pallets and crutch and (b) the power absorbed by the locking and unlocking of the escape wheel. In this respect, the move from the verge, with the pendulum linked directly to the pallets, to the anchor escapement, was a marked improvement. A further advance took place with the perfection of the dead beat escapement which avoids train reversal, with the inevitable errors inherent therein, reduces the effect of the movement on the pendulum and by employing a narrower arc of swing minimizes circular errors.

As mentioned earlier, both Towneley and Tompion[9] were aware of the advantages of the dead beat escapement by 1675-76 and had produced versions of it for the year clocks at Greenwich. Moreover Tompion had fitted a dead beat anchor escapement to Flamsteed's own clock by this time. In view of this, it seems surprising that this escapement had not come into general use prior to 1715, the date when Graham is credited with, but never claimed its invention.

An interesting sidereal and meantime longcase clock by Tompion, circa 1709, (Fig. 9-8) was exhibited by Derek Roberts in June 1986[10] which appeared to have its original deadbeat escape wheel. This seemed to be confirmed by the Report of the Research Laboratory for Archaeology and the History of Art of Oxford University but nothing was proved by their findings.

It was at around this time that Graham joined Tompion and it is unlikely that he would not have known of Tompion's earlier work on the dead beat escapement and indeed possibly worked with him on its development.

Probably Derek Howse's summary of the situation[12] is the best one: *However, though these pioneers (Tompion and Towneley) may have been aiming in the right direction, the fact that the anchor recoil escapement held the field for the next 40 years (from 1675) seems to indicate that they failed to come up with a solution which worked. The writer humbly suggests that Graham should in future be credited, not with the first dead-beat escapement, but with the first successful one.*

The Royal Observatory and Thomas Tompion's Clocks at Greenwich

The birth of the pendulum c.1656 had given enormous impetus to the development of accurate clocks and it came at a time when scientists were moving forward quite rapidly in the field of astronomy and pondering over such problems as whether the earth rotated at a constant speed, something they believed to be true but could not prove and needed to before they could proceed to other matters such as making accurate charts of the stars, the determination of longitude at sea and the provision of accurate tables for the Equation of Time i.e. the difference between Solar and Mean Solar time.

The founding of the Royal Society in London in 1662 was another indication of the rapid advances which were being made in our basic scientific knowledge and the debate and development which was occurring in an era of great academics and scientists such as Robert Hooke, Jonas Moore, Christopher Wren and Isaac Newton.

It is thus scarcely surprising that on June 22, 1675 the King signed a warrant for the building of an observatory at Greenwich (Fig. 2-2), to be designed by Sir Christopher Wren, just eight years after the completion of the one in Paris and three months after Flamsteed (Fig. 2-3) was appointed "Astronomical Observer." Initially, Flamsteed set up an observatory in a turret of the White Tower at the Tower of London but within a remarkably short space of time had moved to Greenwich and indeed was making observations there by September 1675, well before the building was finished. To assist in these observations, he used a longcase clock of his own made by Thomas Tompion with a seconds beating pendulum and, it is thought, maintaining power. Unfortunately the whereabouts of this clock are no longer known. It was installed on the balcony of the Queen's House at Greenwich and was probably the catalyst for the two year clocks which were ordered from Tompion.

Fig. 2-2. **The Royal Observatory, Greenwich. Commissioned by the King on June 22nd 1675 and designed by Sir Christopher Wren.** National Maritime Museum.

It was with his mind on the work which he wanted to carry out at Greenwich and the accuracy of the clocks which would be required for this that Flamsteed wrote to his friend Towneley[13], a gifted astronomer and horologist, as follows: *I am glad to know that the error of your clocks proceeded not from the weather. If you think their motions isochronal I have found new work for them this morning and send you a notion wholly, for aught I know, new to put into experiment. Briefly the Equations of Natural Days are yet in controversy amongst us, and though I have demonstrated in the Diatriba printed with Mr. Horrox remains, and that the Astronomia can only be true, yet it is questionable whether the daily return of any meridian on earth to a fixed star be equal and isochronical at all times of the year.* This clearly refers to the production of accurate Equation of Time tables and the investigation of the isochronicity of the earth.

The result of this letter was the production of two year duration clocks by Thomas Tompion (Figs. 2-4A, B, C, D), to the order of Sir Jonas Moore, (Fig. 2-5) Flamsteed's patron, who also paid for the astronomical instruments. They were fitted with a form of dead beat escapement devised by Towneley, who probably made other important contributions to their design.[14]

Fig. 2-3. **John Flamsteed. The first Astronomer Royal, who was appointed in 1675.** He started making observations at the Tower of London but before the year was out had moved to Greenwich. National Maritime Museum.

Figs. 2-4A-D. **The 17.5" square dials and movements of the two year duration clocks installed in the Octagon Room at Greenwich.** The dials bear the inscription "MOTUS ANNUS." Sir Jonas Moore caused this movement with great care to be thus made A.D. 1676 by Thos Tompion." Both dials would have originally been covered in velvet and thus the dial plates were left unfinished. The clocks were removed from the observatory in 1719 and subsequently had their escapements changed and seconds pendulums fitted.

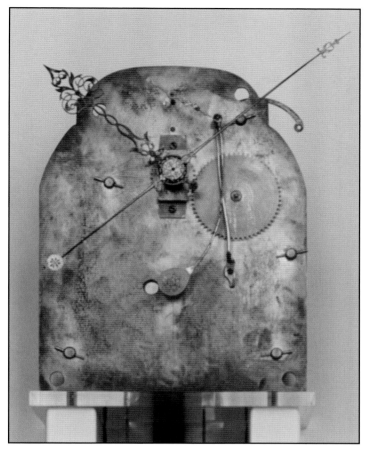

Left and above:
Figs. 2-4A, B. **This is the clock that was until recently owned by the Earl of Leicester and is now at the National Maritime Museum.** The oak case was made up for it in the mid 18th century. The dial is covered by velvet.

Left and above:
Figs. 2-4C, D. **This clock came into the possession of Sir James Lowther and is now in the British Museum.** As shown the velvet is no longer present on the dial and its unfinished state can clearly be seen. Maintaining power was provided, primarily to protect the escapement during winding. This clock has now had the pendulum and escapement reversibly put back to its original design.

Fig. 2-5. **Sir Jonas Moore.** National Maritime Museum.

What is of equal interest is a letter written by Flamsteed to Towneley on 11th Dec 1675 in which he includes a drawing of the escape wheel fitted to his own seconds beating longcase clock made by Tompion and was first referred to in a letter to Towneley dated 2nd March 1675. This clearly shows a form of dead beat escapement (Fig. 2-6). The following remark included in his letter, *but little check is given to the second finger* which was attributed to Tompion, indicates that even at this stage the advantages of the dead beat escapement were appreciated by him.

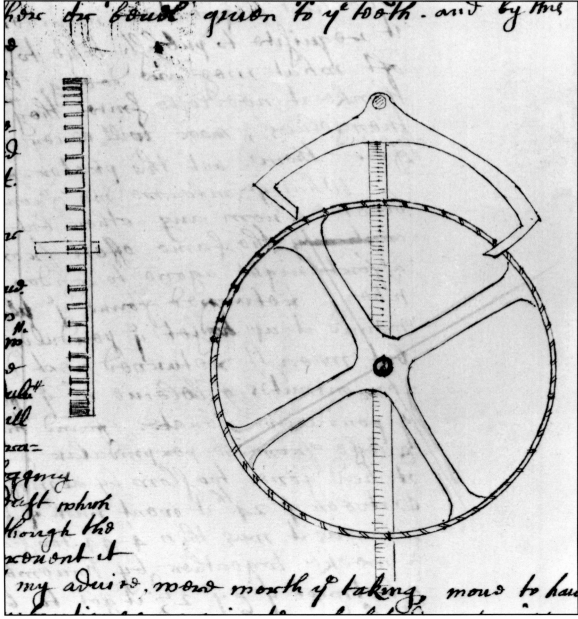

Fig. 2-6. **1675 drawing of Tompion's dead beat anchor escapement.**

To investigate this, the National Maritime Museum asked Richard Good to produce a working model, based on Flamsteed's drawing. To make it work, the pallets had to be narrower and their angle slightly changed, however it then functioned correctly and is illustrated on p.23 of Derek Howse's article.[11]

It was on 7th July 1675 that Tompion's two year clocks were brought to Greenwich and some two months later installed in the Octagon Room (Fig. 2-7). Both had pendulums that beat two seconds some thirteen feet long, of which the details are not known, mounted above the movements, as Towneley had already done on other clocks, and were provided with maintaining power.

Fig. 2-7. **The Octagon Room.** National Maritime Museum.

They were used for recording the eclipses of the Sun and Moon and of Jupiter's satellites and timing occultations of stars and planets by the Moon and investigations concerning the Equation of natural days. Unfortunately the performance of the clocks in the first eighteen to twenty-four months proved unsatisfactory and in the end Tompion removed Towneley's pallets and produced a modified version, which were also deadbeat and once again the clocks were set in motion. Initially there were teething problems but within six months they had settled down and appeared to give no further problems, no stoppages being recorded until one was cleaned some four years later. It is assumed that they continued to give satisfactory service until, following Flamsteed's death in 1719, they were removed from the Observatory.

A surprising feature of the clock's movements and pendulums is that although behind the paneling, they were not encased in any other way, which must have made them subject to dust and the influence of any draughts and indeed in March 1687 it is recorded that they both stopped because of high winds outside.

Sadly no records seem to have been kept by Flamsteed of the performance of the clocks once they had settled down; however it must have been reasonably satisfactory as Flamsteed wrote to Sir Jonas Moore in March 1678 stating that he had proved the isochronicity of the earth's revolutions, although nothing was published. Flamsteed's equation tables were subsequently produced which assisted in much other useful research.

Although no formal records of Tompion's clocks

Fig. 2-8. **Tompion's Degree Clock.** This was supplied to Flamsteed in 1691 and gives sidereal time measured in degrees, minutes and seconds of arc. The arc minutes and arc seconds are shown on the chapter rings and the degrees by a disc seen through an aperture reading from 1-36 with each division representing 10°. The clock has a 2/3$^{rd.}$ second beating pendulum and a duration of one month. It may be seen in the National Maritime Museum, Greenwich.

are known, one cannot help feeling that Flamsteed would almost certainly have mentioned them in correspondence but to whom and when and whether these letters still exist is another matter. It could be that they were referred to in communications between Flamsteed and his patron Sir Jonas Moore, much of which still exists, but is in Latin.

A third clock was supplied to Greenwich, as is indicated in a picture of the Octagon Room where a third dial is shown, above which may be seen structural alterations, but this subject is outside the scope of this book.

One other clock should be mentioned here, Tompion's degree clock (Fig. 2-8) which he supplied to Flamsteed in 1691 and is on display in The Old Royal Observatory, National Maritime Museum. This shows sidereal time but measured in degrees, minutes and seconds of arc instead of hours, minutes and seconds of time. It is month going and has a $2/3^{rds}$ seconds beating pendulum.

The original specification was that the middle and lower dials should show minutes and seconds as on a conventional clock but that the upper index should be arranged to make one revolution for every thirty-six of the middle one; however Tompion modified this, putting the "seconds" hand reading arc minutes and arc seconds at the top of the dial, driven directly by the escape wheel and the hour hand was replaced by a disk, viewed through an aperture in the dial, reading from one to thirty-six, each division representing 10°.

To gain a much fuller account of the Observatory and its clocks one can do no better than read the superbly researched accounts by Derek Howse.[11, 12]

References

[1] Hurst, M. *The first twelve years of the English Pendulum Clock*. Antiquarian Horology. June 1969. pp. 146-156.

[2] Edwardes, E.L. & Dobson, R.D. *The Fromanteels and the Pendulum Clock*. Antiquarian Horology. Sept. 1983. pp. 250-263.

[3] M.H, Robert. *The Retardation of a Pendulum caused by an increase in its arc of oscillation*. Reproduced in Saunier C. Treatise on modern Horology 1861. Translated from the French, 1975. R.W. Foyle, London. p. 692.

[4] *The Invention of the Anchor Escapement*. Antiquarian Horology, June 1971. pp. 225-228.

[5] Beeson, C.F.C. *A history of Wadham College Clock*. Antiquarian Horology, June 1957. pp. 47-50.

[6] Turner, A.J. *Christopher Wren and the Wadham Clock*. Antiquarian Horology, June 1971. pp. 229-230.

[7] Greenwood, C. *A Joseph Knibb Longcase Clock with early anchor escapement*. Antiquarian Horology. Spring 1988. pp. 259-263.

[8] Derham, W. *The Artificial Clockmaker 1696*. James Knapton, London.

[9] Smith, A. *A reconstruction of the Tompion/Towneley/Flamsteed "Great Clocks" at Greenwich*. Antiquarian Horology. Dec. 1999. pp. 185-190.

[10] Roberts, D. *Precision Pendulum Clocks*. Private Publication 1986.

[11] Howse, D. *The Tompion Clocks at Greenwich and the Dead-beat escapement, Part 1 1675-1678*. Antiquarian Horology. Dec. 1970. pp.18-34.

[12] Howse, D. *The Tompion Clocks at Greenwich and the Dead-beat escapement, Part 2 1678-1691 with an appendix by Beresford Hutchinson*. Antiquarian Horology, March 1971. pp.114-133.

[13] Webster, C. *Richard Towneley 1629 - 1707 and the Towneley Group*. Trans Hist. Soc Lancashire & Cheshire, Vol. 118. pp. 51-76 1966.

[14] Smith, A. *A reconstruction of the Tompion/Towneley/Flamsteed "Great Clocks" at Greenwich*. Antiquarian Horology. Dec. 1999. pp. 185-190.

Chapter 3
Sidereal, Solar, Mean Solar and the Equation of Time, Local and Greenwich Mean Time

BY DEREK ROBERTS

Solar and Mean Solar and the Equation of Time

The solar day is the time taken between two consecutive passages of the sun across the meridian of the place of observation and is recorded from mid-day to mid-day. It was known by Islamic astronomers in the middle ages that the lengths of these days varied throughout the year but it was only in the seventeenth century that this started to become an inconvenience; however at that time no clock was available which was precise enough to measure these differences with any degree of accuracy.

To overcome the problem of the varying length of the day, Mean Solar Time, an expression usually shortened to Mean Time, which was also sometimes referred to as equal time, was evolved which gave each day throughout the year exactly the same duration. This was determined by averaging out the varying lengths of the solar days. The difference between apparent solar time and mean solar time on any given day is known as the Equation of Time, also in earlier times called the Equation of Natural Days. The variations in solar time are due to the earth's inclination on its axis to the ecliptic and its passage in an ellipse around the sun, with the earth's speed varying, being greatest when it is nearest the sun.

The value of having accurate tables (Fig. 3-1) giving the difference between Mean and Apparent Solar Time was very great as not only was it of use to astronomers but to anyone who used a watch or clock, particularly if it was a public one on which the local populace relied. By looking at the sun-dial, which on the better examples could be read off to within a minute and then adding on or taking off the amount indicated by the equation tables, the clock could be set to time, and indeed on some sun-dials this information was recorded on them (Fig. 3-2A, B).

A TABLE of the Equation of natural Days, ſhewing what Time it ought to be by the Clock when the Sun is on the Meridian.

The firſt after Leap-year.

Days	January			Dif.	February			Dif.	March			Dif.	April			Dif.	May			Dif.	June			Dif.
	H.	M.	S.	S.	H.	M.	S	S.	H.	M.	S.	S.	H.	M.	S.	S.	H.	M.	S.	S.	H.	M.	S.	S.
1	12	4	21	Inc.	12	14	11	Inc.	12	12	39	Dec.	12	3	52	Dec.	11	56	49	Dec.	11	57	18	Inc
2	12	4	49	28	12	14	18	7	12	12	26	13	12	3	34	18	11	56	41	8	11	57	27	9
3	12	5	16	27	12	14	24	6	12	12	13	13	12	3	16	18	11	56	34	7	11	57	36	9
4	12	5	43	27	12	14	29	5	12	12	0	13	12	2	58	18	11	56	28	6	11	57	46	10
5	12	6	10	27	12	14	34	5	12	11	46	14	12	2	40	18	11	56	22	6	11	57	56	10
				27				4				14				18				5				10
6	12	6	37		12	14	38		12	11	32		12	2	22		11	56	17		11	58	6	
7	12	7	3	26	12	14	41	3	12	11	17	15	12	2	4	18	11	56	13	4	11	58	17	11
8	12	7	28	25	12	14	43	2	12	11	2	15	12	1	47	17	11	56	9	4	11	58	28	11
9	12	7	53	25	12	14	44	1	12	10	46	16	12	1	30	17	11	56	5	4	11	58	40	12
10	12	8	17	24	12	14	44	0	12	10	30	16	12	1	13	17	11	56	2	3	11	58	52	12
				24				Dec.				16				16				2				12
11	12	8	41		12	14	44		12	10	14		12	0	57		11	56	0		11	59	4	
12	12	9	4	23	12	14	43	1	12	9	57	17	12	0	41	16	11	55	58	2	11	59	16	12
13	12	9	26	22	12	14	41	2	12	9	40	17	12	0	25	16	11	55	57	1	11	59	28	12
14	12	9	48	22	12	14	38	3	12	9	23	17	12	0	9	16	11	55	56	1	11	59	40	12
15	12	10	9	21	12	14	35	3	12	9	6	17	11	59	54	15	11	55	56	Inc.	11	59	53	13
				21				4				18				15				1				13
16	12	10	30		12	14	31		12	8	48		11	59	39		11	55	57		12	0	6	
17	12	10	50	20	12	14	26	5	12	8	30	18	11	59	25	14	11	55	58	1	12	0	19	13
18	12	11	9	19	12	14	20	6	12	8	12	18	11	59	11	14	11	56	0	2	12	0	32	13
19	12	11	27	18	12	14	14	6	12	7	54	18	11	58	57	14	11	56	2	2	12	0	45	13
20	12	11	44	17	12	14	7	7	12	7	36	18	11	58	44	13	11	56	5	3	12	0	58	13
				17				7				18				13				3				13
21	12	12	1		12	14	0		12	7	18		11	58	31		11	56	8		12	1	11	
22	12	12	17	16	12	13	52	8	12	6	59	19	11	58	19	12	11	56	12	4	12	1	24	13
23	12	12	32	15	12	13	43	9	12	6	40	19	11	58	7	12	11	56	16	4	12	1	37	13
24	12	12	46	14	12	13	34	9	12	6	21	19	11	57	55	12	11	56	21	5	12	1	50	13
25	12	12	59	13	12	13	24	10	12	6	2	19	11	57	44	11	11	56	27	6	12	2	3	13
				13				10				19				10				6				13
26	12	13	12		12	13	14		12	5	43		11	57	34		11	56	33		12	2	16	
27	12	13	24	12	12	13	3	11	12	5	24	19	11	57	24	10	11	56	39	6	12	2	28	12
28	12	13	35	11	12	12	51	12	12	5	5	19	11	57	14	10	11	56	46	7	12	2	4[illegible]	12
29	12	13	45	10				12	12	4	46	19	11	57	5	9	11	56	53	7	12	2	5[illegible]	12
30	12	13	54	9					12	4	28	18	11	56	57	8	11	57	1	8	12	3	4	12
				9								18				8				8				11
31	12	14	3	8					12	4	10	18					11	57	9	9				
	Incr. 9′ 42″				Incr. 0′ 33″ Decr. 1 53				Decr. 8′ 29″				Decr. 6′ 55″				Decr. 0′ 53″ Incr. 1 12				Incr. 5′ 46″			

Fig. 3-1. **An Equation of Time table for the period Jan - June taken from Ferguson's Astronomy**. Besides giving the difference between solar and mean solar time it also shows the change in seconds from one day to another.

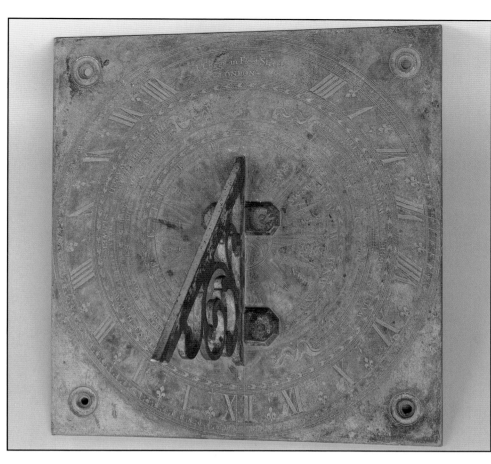

Figs. 3-2A, B. **A garden sun-dial by I. Coggs in Fleet Street, London. c.1740** which can be read to the nearest minute and also gives the Equation of Time (Watch Slower-Watch Faster) throughout the year. At this period the term "watch" was used both for watches and clocks when related to recording time.

With the advent of the long pendulum and anchor escapement the accuracy of clocks increased dramatically and so that their owners could check and regulate them, Equation of Time tables were quite often pasted inside the trunk door (Fig. 3-3), Tompion for instance commissioning his own tables for that purpose in 1683.

Fig. 3-3. **Equation of Time Tables, pasted inside the door of a longcase clock by Quare & Horseman dated 1733.** They also contain instructions for taking care of the clock.

One of Flamsteed's main aims in his early days at the observatory was to research and compile an accurate Equation of Time table. Christian Huygens had already produced one in 1665 and by 1669 it had been translated into English and published in the Philosophical Transactions for that year.[1]

In 1666 John Willis produced his *Hypothesis about the Flux and Reflux of the Sea*[2] where he stated that there were at least two causes for the inequality of the natural day. The first cause was the variation in the Earth's annual orbit around the sun, given by the tables of the sun's annual motion. The second cause was the tilt of its axis to the plane of the Earth's orbit, given by the table of the sun's right ascension.

In 1671 John Wallis was preparing the Horrox papers, which came from Towneley Hall, for publication and it was around this time that Flamsteed let him have his equation tables. These he added to the Horrox papers as an appendix and they were published in 1673. They commence with the position of the sun at the spring equinox and are thus related to the sun's position in the zodiac.

In 1676 an Equation of Time table was published in the *Royal Almanac* using Flamsteed's earlier figures. This was based on the Julian calendar rather than the sun's position in the zodiac. Unfortunately, in the transference from the Gregorian calendar ten days were added instead of subtracted.

In 1670 Flamsteed's tables for the following four years were included in the second edition of the *Artificial Clockmaker* by William Derham and in the following years Equation of Time tables were published on a fairly regular basis using various different sources for their information.

Much fuller and excellently researched accounts of the origin of Equation of Time tables and John Flamsteed's involvement in these are included in articles by Tony Kitto[3] and Derek Howse.[4]

Methods of Indicating Solar Time

To simplify still further the use or display of the difference between mean and apparent solar time, which varies by as much as fifteen minutes during the year, various other methods were devised, some of which are illustrated here.

Joseph Williamson claimed in a letter which was published in *Philosophical Transactions* at the end of 1719, that he asserts his Right to the curious and useful invention of making clocks to keep time with the sun's Apparent Motion which probably refers to his Solar Time Clocks. He does go on to discuss a clock by Quare which he claims to have made, which was supplied to King Charles II of Spain, which showed both mean and apparent solar time, making use of a cam to lengthen or shorten the effective length of the pendulum and others which he made for Quare in the same way; however it seems likely that he only made the equation part and not the whole clock. There is no indication that this was true of Tompion, as all the work on his equation clocks is definitely his own as it is also with some other makers.

The confusion could arise from the fact that Williamson was indeed probably responsible for all the clocks showing apparent solar time which were made at this period, but this is another matter. They will be dealt with a little later in the chapter.

The main ways which were used to display mean and apparent solar time or the difference between them, known as the Equation of Time, are (a) the use of a large annual calendar disc on which the figures for the Equation of Time are also given (Figs. 3-4 and 3-5); (b) the provision of a single extra hand, controlled by a cam, indicating the Equation of Time such as that seen in the arch of Tompion's clock (Fig. 3-6); (c) the use of two dials and two pendulums, the effective length of one of which is varied throughout the year to give solar time (whereas on Quare and Horseman's clock, Fig. 3-7, the two dials are shown alongside each other; on the one by Williamson they are back to back); (d) the use of two chapter rings, one stationary, giving equal or mean-time and the other moving, indicating solar time. The difference in the readings on the two chapter rings will give the Equation of Time (Fig. 3-8); (e) a manually set equation hand as on the clock by Thiout L'Ainé (Fig. 3-9) in which the hand is set by reference to an Equation of Time table on the trunk door and (f) the use of two minute hands, (Fig 3-10A, B) one displaying mean and the other solar time, with the difference between the two giving the Equation of Time. The position of the two hands relative to each other is determined by a cam which rotates once a year and advances or retards the second hand by the correct amount.

Fig. 3-4. **A fine lacquer longcase clock by John Topping, c. 1725 with year calendar in the lower half of the dial center that also gives the Equation of Time.** Note the delicacy of the hands used to reduce the masking of the year disc to a minimum. A rare feature is the display of seconds in the arch.

Figs. 3-5A, B. **A month duration walnut longcase clock by Daniel Delander with year calendar and Equation of Time, titled "Equation of Natural Days" in the arch.**

Fig. 3-6. **Thomas Tompion's clock with year calendar and Equation of Time, presented to the Pump Room, Bath.**

Fig. 3-7. **Quare & Horseman's double dialed clock with two pendulums, now in the British Museum.** This is, in effect, two separate clocks, with the pendulum of one adjusted for Sidereal Time so that this may be shown on the dial, and the other giving mean-time.

Fig. 3-8A, B. **Thomas Tompion.** A year duration walnut longcase clock made for William III at the end of the 17th century. The single train movement has the escapement inverted. The main chapter ring has 2 x XII hour numerals with minutes on its outer edge. This is inscribed equal (mean time). Outside this is a second narrower chapter ring bearing the numerals 0-60 twice for the minutes, the minute hand going round every two hours. The difference in the reading between the two chapters gives the Equation of Time. In the arch are two apertures, the upper one showing the signs of the zodiac and the sun's position in the ecliptic and the lower one having an annual calendar. The mechanism for advancing or retarding the outside chapter ring may be seen in Fig. 3-8B. Royal Collection.

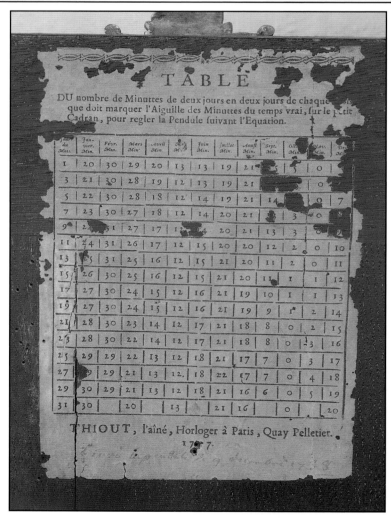

Figs. 3-9A, B, C. **Thiout's manual equation clock that is set by referring to the table inside the door.** This is described in Thiout's *Traité de l'Horlogierie* Vol. II 1741. pp. 279-282 and plate 28.

Above and following page:
Figs. 3-10A, B. **Bailly, Paris.** A fine longcase regulator with remontoire above the main movement. Besides seconds, hour and date hands there are individual hands for both mean and solar time with the difference between these giving the Equation of Time. The cam in the center of the year wheel with a roller resting on its edge advances or retards the solar hand throughout the year so that it always indicates solar time correctly.

34

Figs. 3-11A, B. **B. Parr, Grantham. c. 1790.** A highly unusual longcase clock with an aperture above 6 o'clock for the annual calendar and above this a month calendar shown in a sector with a gilded "fly back" hand. Below 12 o'clock is a hand showing the Equation of Time graduated 15-0-15. This is moved by means of the year calendar disc which, although it appears circular, is not, its minor deviation from the round being magnified up by a pulley running on its edge and various levers.

Fig. 3-12. **Edward Cockey, Warminster. c. 1705.** This clock is included here to give some idea of the complexity of the dials produced at around this period and also a little later in England and at least a century earlier in Germany.

The dial plate: The dial plate is 20" square and carries a twenty-four hour chapter ring. Within this is set a dish which rotates once in 24 hours. This dish is painted to represent the day and night skies and carries a gilt representation of the sun, not seen in this picture. The dish revolves behind a painted fixed shield which fully obscures the celestial scene between the hours of 8.15 p.m. and 3.45 a.m. Appearing from each end of this shield are blued steel shutters which rise and fall according to the season and ensure that the sun rises and sets at the correct time each day.

The sunrise/sunset shutters are controlled through arms which ride on large cams which are each attached to one of two 9" diameter wheels with 365 teeth which rotate once a year. Fitted to the bottom shield is a 10" diameter brass ring decorated with wheatear engraving which acts as a guide to the inner edges of the sunrise/sunset shutters. Within this fixed ring is a blued steel wavy edged ring which rotates with the celestial dish and attached to this is a cast gilt figure of Father Time. His hand indicates the hour, whilst the minutes are shown by the long gilt center sweep hand.

Within the blue wavy ring are set three concentric silvered rings. The outer one is a year calendar, the next represents the ecliptic divided into the 360° of the Zodiac, and, within that, is shown the 29-1/2 days of one lunation.

The calendar and zodiac scales are fixed together and are read from the pierced gilt pointer fitted to the blue ring at 90° from Father Time. These two rings have been moved eleven days relative to each other during the eighteenth century to correct for the change, in 1752, to Gregorian Calendar.

The age of the moon, read from the inner of the three rings, is indicated by the right foot of Father Time. Set within the lunar calendar in a blued plate with gilt stars is a moon ball and this rotates about a radial axis to show its phases.

In the center of the dial, set on a silvered plate engraved to show clouds and rain, is a representation of the earth. It is interesting to note that, viewed from the earth, the sun and moon always maintain their correct relative position in the Zodiac.

Joseph Williamson's Solar Time Clocks

These ingenious clocks, which show apparent solar as opposed to mean solar time, were made over a period of maybe ten years with the design gradually evolving during this period. As they displayed solar time, their accuracy could be compared directly with the time shown on a sundial.

The difficulty which Williamson faced was to produce a clock where the timekeeping was constantly changing, i.e. the clock had to be continuously speeded up or slowed down and always by a precise amount. This was achieved by using a cam which rotates once a year and as it does so raises or lowers the top block of the suspension spring a very small amount each day. As the steel strip passes through a slit in a brass block, its effective length is constantly varied. The same principle was to be used by Brocot over one hundred years later to provide fast/slow regulation on French mantel clocks.

Probably the earliest dial layout used by Joseph Williamson was that seen in Tom Robinson's book[5] in which the annual calendar with the signs of the zodiac is in the arch; there is a seconds' ring in the conventional position below 12 o'clock and Williamson's inscription *Horae indicantor. Apparentes Involutis Aequationibus Joseph Williamson Londini* (the apparent (solar) hours are shown by means of complex equation work) in the lower half of the dial center.

The clock illustrated in Fig. 3-13A, B, C, which is probably of a slightly later date, is an advance on this in that, by moving the plaque to the top of the dial center, space has been made below to show the moon and its age; however this has meant sacrificing the seconds' ring.

Right and following page:
Figs. 3-13A, B, C. **Joseph Williamson, London. c. 1720.** In the rear view of the movement may be seen the spiral drive, via a contrate wheel below, to the year calendar with the signs of the zodiac in the arch and the equation disc mounted above and behind the backplate. This appears almost circular. A pivoted cranked lever rests on its edge whilst from the other end the pendulum is suspended and is thus raised or lowered as the cam rotates, which, as the suspension spring passes through a narrow slot, varies its effective length and thus its timekeeping throughout the year to match the variations in the length of the solar day.

In the arch of the 12" dial is a large ring which indicates, from the center outwards, the signs of the zodiac, the months and the days of the month. In the lower half of the dial center are separate quite large apertures displaying the moon and indicating its age.

The final dial layout used by Williamson is probably that seen in Fig. 3-14A, B, in which seconds have been reinstated by using a center sweep hand and the display of the moon has been increased in size.

Figs. 3-14A, B. **Joseph Williamson, London. c. 1725.** This probably represents Williamson's final development of his dial. It is similar to that seen in Figs. 3-13A, B, C but a center sweep seconds' hand has been added and the apertures for the moon and lunar calendar enlarged.

Sidereal Time

The time required for the earth to make a complete revolution is known as a sidereal day and as the earth keeps almost perfect time this is virtually a constant. It is measured from mid-day and can be recorded by stars known as "time stars" which are infinitely remote and therefore make the relative movement of the earth around the sun insignificant. By using a telescope known as a transit instrument that is mounted on a fixed N/S axis (its meridian), the time taken between the stars disappearing from view for the first and second time, can be recorded. To minimize any errors it is usual to take the readings from several stars on any one night and average out the results.

Harrison used a time star to check the accuracy of his regulators, noting the time when the star, aligned with the window frame, disappeared from view behind a distant building and Breguet refined this technique[6] by using a rotating six armed spider with a small black disc at the end of each limb, one of which will cross the observer's field of vision every 0.1 second and enable him to determine the time of the transit to this degree of accuracy.

Sidereal time is divided up into hours, minutes and seconds as is "mean (solar) time" and these are the standard units of time used in astronomy. Thus an astronomical regulator in an observatory will normally be set up to show sidereal time, the day being 3 minutes 55.91seconds shorter than the mean solar day which equates to 9.83 seconds an hour. It will thus have a slightly shorter pendulum than a mean time clock.

Occasionally a regulator is to be seen which shows both Sidereal and Mean Time, such a regulator by Brockbanks being shown in Fig. 3-15A, B, C, D, E and Margetts' clock in Fig. 3-16A, B, C, D. The pendulum is adjusted to give sidereal time and a complex series of gears of very high count is then required to give the correct ratio between mean and sidereal time. Probably the finest sidereal and meantime regulator the author has seen is that by Cooke (Fig. 18-20) who made a small series.

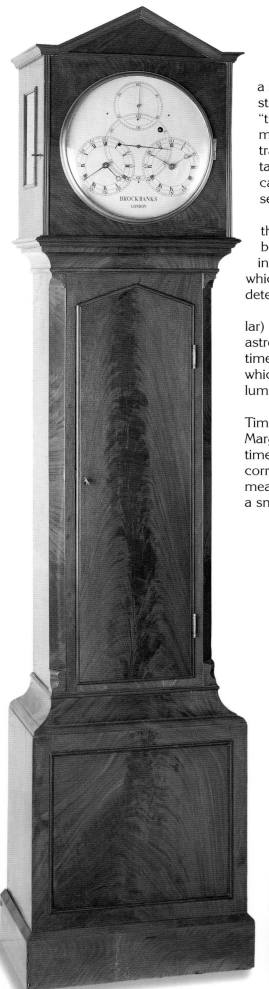

Fig. 3-15A

Figs. 3-15A, B, C, D, E. **Brockbanks, London. A Sidereal and Mean Time Regulator.** This early 19th century longcase regulator has a silvered brass dial with sidereal time, employing blued steel hands, being displayed on the left (minutes and hours) and in the center (seconds) whilst mean time is indicated at the top and bottom right.

The regulator employs a pendulum beating sidereal seconds with a complex gear train effecting the necessary change to mean time. One mean solar day is 24 hrs. 3 mins. and 56.55536 seconds in sidereal time giving a ratio required by the gear train of as close as possible to 1:1.0027379093. Vines (Fig. 18-12) calculated a train so close that the error was only one second in over five years. The train in Margett's regulators was also very accurate with an error of just 1.8 seconds a year.

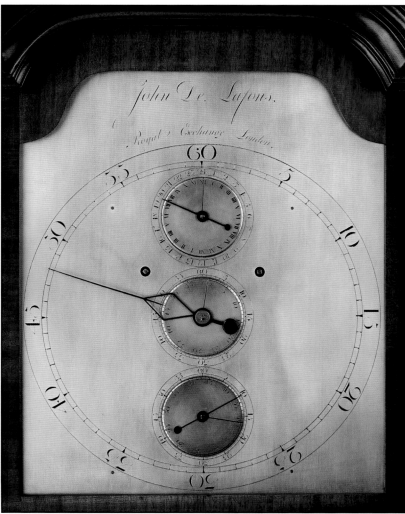

Figs. 3-16A-D. **George Margetts, London.** A late 18th century sidereal and meantime regulator by this maker but signed John De Lafons, Royal Exchange, London, for whom he presumably made it. George Margetts and his work is described elsewhere in this book but this clock is included here to give some indication of the complexity and fineness of the wheelwork required to obtain an accurate conversion from sidereal to mean time. Margetts' train of gears resulted in the mean time being less than two seconds slow at the end of one year. He was a fine mathematician.

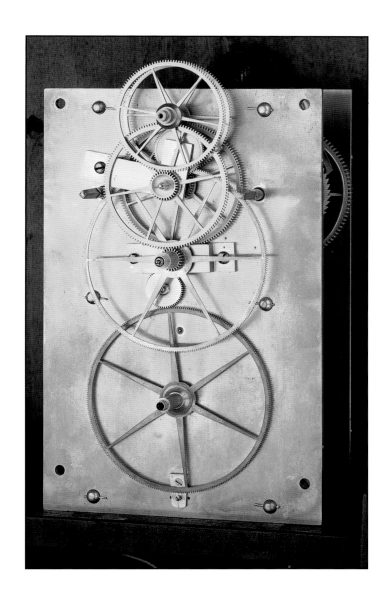

Local and Greenwich Mean Time

Although it was in 1675 that the Royal Observatory at Greenwich was founded and Tompion's year duration clocks set going there very shortly afterwards, which were used to prove that the earth rotated at a constant speed, it was not until some two hundred years later that time was standardized throughout the United Kingdom. Prior to this each city or district based its time initially on local solar time, i.e. the time measured by the passage of the sun immediately overhead in that city at mid-day. As this varied throughout the year and was therefore inconvenient, by the end of the eighteenth century local mean solar time had been adopted; however this still left a difference of roughly thirty minutes between the extreme east and west of the country.

The situation at sea was different as the difference in mean time between Greenwich and aboard ship had to be known quite precisely if accurate navigation was to be achieved. This had become a reality with the invention of the marine chronometer.

The two major forces which were to bring about the standardization of Greenwich Mean Time throughout the land were the increasing use of stage coaches in the last part of the eighteenth century, which kept remarkably accurate time both for passenger transport and to carry the Royal Mail. The latter was particularly important as they had to synchronize watches for the interchange of letters for different parts of the country. It was obviously impractical for the coachman to constantly change his watch as he continued on his journey and the only sensible solution was to display Greenwich Mean Time at each of the points of arrival and departure to avoid passengers missing the coach.

This problem was greatly exacerbated by the appearance on the scene of the steam locomotive with its much higher speeds and the imposition of strict timetables following the invention of the electric telegraph. Fortunately this coincided with the increasing ability to transmit the time throughout the Kingdom.

One of the first steps to standardization was the installation of Time Balls (Fig. 3-17A, B, C) which were dropped at exactly one o'clock, this being chosen to avoid 12 o'clock which was the time when all ships chronometers were wound. One of the first of these was installed at Greenwich in 1833 and it still falls there today. Others gradually followed in various other ports[7] and came to be used, not just to set chronometers, but also clocks used by the general populace. Pocket chronometers were also employed to disseminate accurate time. It was at this period that railway timetables were issued which gave both local and Greenwich Mean Time[8] and pamphlets produced showing the time difference between the main towns and cities of the Kingdom.

Figs. 3-17A, B, C. **The Time Ball Tower at Deal, installed c. 1855.**[7] Deal was at one time a busy channel port, being used both as a supply yard for the Royal Navy and also as a safe anchorage in stormy weather.

To enable ships to set their chronometers, which was essential for accurate navigation, the first time-ball, which was dropped at exactly 1pm, was installed at Greenwich in 1833. That at Deal was commissioned some 22 years later. In Figs. 3-17B, C may be seen the controlling mechanism. Fig. 3-17B shows the central pillar of the building on which the capstan is mounted to raise the ball; the return signal mechanism and the modern quartz controlled timer. In Fig. 3-17C is the reconstructed "pecker-box" and Wheatstone galvanometer which received the signal from Greenwich via Deal Station; the Shepherd Clock and the Rating Book. Above the "pecker-box" is a small push switch which was thought to be used to drop the time ball manually if a malfunction occurred.

The construction of "Big Ben" in London in 1859 gave a time standard for the city because of its remarkable accuracy. One interesting fact was the issuing of charts showing the correction that had to be made, if one were relying on listening to its strike (Fig. 3-18), because of the time the sound would take to reach you.

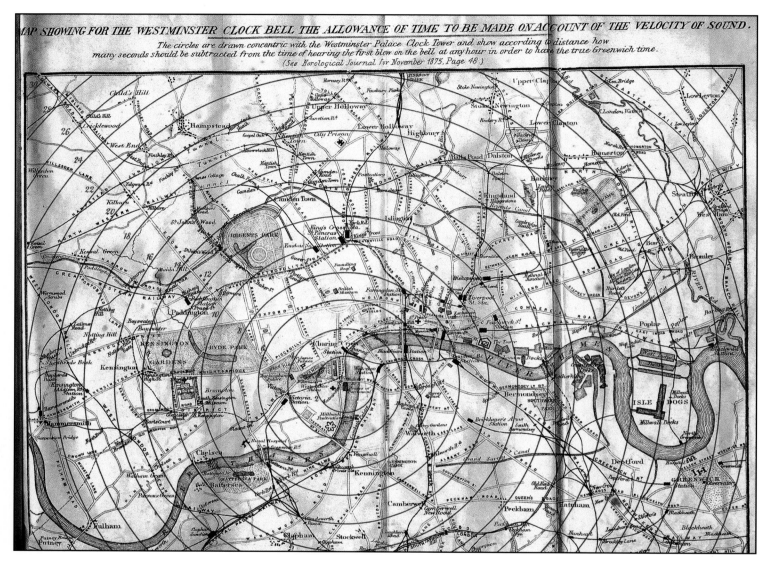

Fig. 3-18. A map showing the allowance that had to be made for the time taken by the sound to reach you from "Big Ben." Taken from the Horological Journal of Nov. 1875.[8]

The installation of Shepherd's Galvano-Magnetic Clock System at Greenwich[9] in 1852 was another leap forward as it enabled accurate time to be transmitted, not just throughout the Observatory but, via the telegraph lines of the South Eastern Railway to stations throughout their region and this rapidly spread all over the country. It was around this time that tables started to appear (Fig. 3-19) showing the difference in time between various cities and towns both in this country and overseas and guidance was given as to which cities throughout the country kept Greenwich and which local mean time (Fig. 3-20).

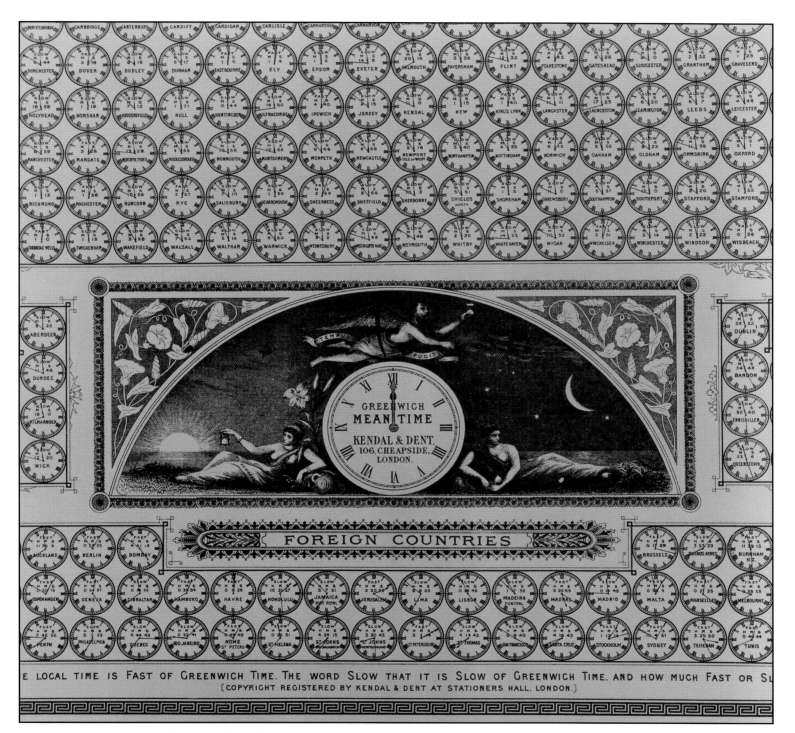

Fig. 3-19. Picture by Kendall & Dent showing the time at various towns throughout the U.K. and also overseas.

Fig. 3-20. A map, published by Henry Ellis and Son in 1852 showing which towns used local and which Greenwich Mean Time.

An interesting regulator (Fig. 3-21A, B, C), which is believed to have been owned by the Harbour Master of Liverpool c. 1850, illustrates the problems which could arise. It shows both the local time at Liverpool and also Greenwich Mean Time.

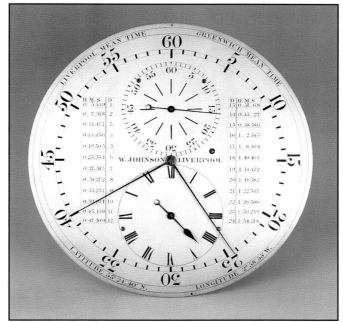

Figs. 3-21A, B, C. **W. Johnson, Liverpool. An interesting regulator giving both Liverpool and Greenwich Mean Time and having Sidereal tables engraved on the dial. c. 1850.** This regulator which, by repute, belonged to the Harbour Master of Liverpool, is the only one we have encountered giving Mean Time both in London (Greenwich) and Liverpool.

To put this in context it must be remembered that it was made approximately 25 years before Greenwich Mean Time was universally adopted. The time differential based on the relative longitudes of the two places would have been approximately 12 minutes, an extremely significant amount so far as navigation is concerned.

The 14" circular silvered brass dial is of conventional regulator layout, but with two center sweep minute hands; one for Greenwich and the other for Liverpool Mean Time. There is a seconds' ring below 12 o'clock and hours are shown at the bottom of the dial. Engraved around the periphery at the top of the dial is, on the left, "Liverpool Mean Time" and on the right "Greenwich Mean Time." At the bottom are given the latitude [53° 24' 40" N] and the longitude [2° 58' 55" W] for Liverpool.

By the 1860s the majority of the country had accepted Greenwich Mean Time and in 1880 the "Definition of Time Act" was brought in stating that in all legal documents the time referred to shall be Greenwich Mean Time.

Some four years later at the international conference in Washington, it was agreed that the International Time Standard, on which all charts would be based, would be the axis of Airy's Transit Instrument at Greenwich.

References

[1] Phil. Trans; Vol IV No. 47 May 10 1669, 937.
[2] Willis, J. Phil Trans. Vol.1 No. 16 Aug 6 1666, 263
[3] Kitto, T. *John Flamsteed, Richard Towneley and the Equation of Time*. Antiquarian Horology. December 1999. pp. 180-184.
[4] Howse, D. *The Tompion Clocks at Greenwich and the Dead-Beat Escapement*. Antiquarian Horology. December 1970. pp. 28-34.
[5] Robinson, T. *The Longcase Clock*. Antique Collectors' Club. p. 180. Fig. 8-31.
[6] Daniels, G. *The Art of Breguet*. p.241. Fig. 270. Sotheby Parke Bernet. London & New York 1975.
[7] Mallinder, D.S. *A time ball at Deal*. Clocks. Vol. 22, 10th October, 1999. pp. 26-28.
[8] Davies, A.C. *The Adoption of Standard Time and the Evolution of Synchronised Timekeeping*. Antiquarian Horology. Spring 1979. pp. 284-289.
[9] Hutchinson, B. *Greenwich Mean Time and Longitude Zero*. Horological Journal. May 1984. pp. 13-16.

Chapter 4
Factors Affecting the Isochronicity of a Pendulum

BY DEREK ROBERTS

The birth of the pendulum, and a little later, the anchor escapement and royal seconds beating pendulum, brought about two dramatic improvements in the accuracy of timekeeping. It was not to be long before it was realized that the pendulum was the all important governing factor in accurate timekeeping, and that anything, including the movement, which disturbed it in any way would affect the clock's timekeeping.

Circular Error

The first factor to be considered was circular error. When Huygens analyzed Galileo's work on the isochronicity of the pendulum he found that Galileo's observation that the time a pendulum took to swing from one side to the other was always the same, no matter how wide the arc of swing, was only accurate if that swing passed through a somewhat steeper path than an arc known as a "cycloidal curve." To overcome this error, which could be appreciable when the verge escapement was used with its relatively wide arc of swing, he suggested the use of "cycloidal cheeks" (Fig. 4-1) placed on either side of the suspension spring to modify the arc of swing. This was used in many early pendulum clocks; however it was never generally adopted and became of far less importance with the advent of the anchor escapement and the relatively narrow arc required for this. However, the importance of circular error is emphasized by the figures produced by M. H. Robert[1], whose tables show that the loss of time involved per day for a pendulum swinging free in an arc of 1° is 1.55 seconds, which increases to 14.8 seconds at 3°; 1.05 minutes at 6° and 24 minutes at 30°. All this, of course, is of no importance if the arc of swing remains constant, but for a variety of reasons this is unlikely to be the case, particularly over a long period of time.

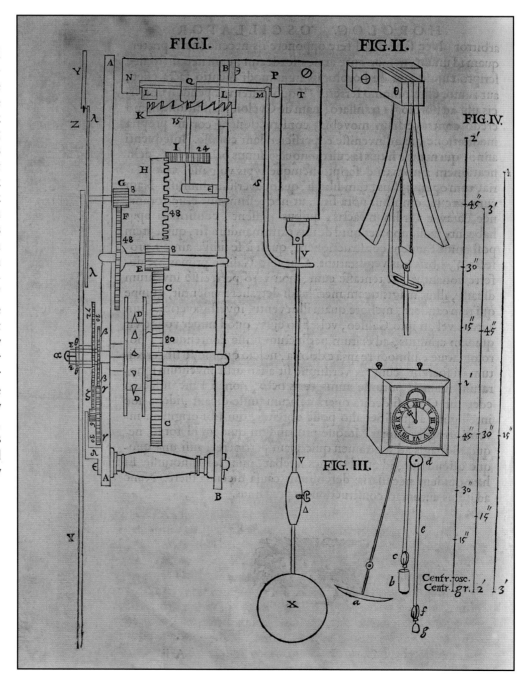

Fig. 4-1. **Cycloidal Cheeks as suggested by Huygens.** This followed his observation that Galileo's discovery that a pendulum always takes the same time to swing from one side to the other, no matter how wide it is, is only true if it moves in a somewhat steeper path than an arc known as a cycloidal curve.

Thermal Compensation

As the accuracy of clocks improved their owners became increasingly aware of changes in timekeeping with temperature and, in particular, between summer and winter as their pendulums expanded or contracted.

George Graham

Scientists had been aware of the varying expansion of different metals with rising temperature since the mid seventeenth century and by the early 1700s clockmakers such as George Graham were investigating and measuring these changes to see if they could devise a pendulum the effective length of which stayed constant. He had thought of using one metal for the rod and another which had a much higher coefficient of expansion for the bob, but found that the coefficients of the common metals such as brass and iron were too close to each other for this purpose. He was well aware however, of the expansion of mercury, having had a mercury thermometer at least by 1712, but had assumed that, being a dense metal, it would have a relatively low rate of expansion. On measuring it he found its coefficient of expansion was very high and thus by 1722 his mercurial pendulum had been born. It was the first truly compensated pendulum to be devised, which he reported to the Royal Society in 1726.[2]

This was to stay in use for the best part of two hundred years and will be expanded on a little later, being refined by Dent, Frodsham and Riefler amongst others.

John Harrison and the Gridiron

The next major contribution to thermal compensation was the gridiron pendulum first made by Harrison c. 1726. Having assessed the relative expansions of steel and brass at roughly 3:5 he realized that if a five foot steel rod expanding down could be opposed by a three foot brass rod expanding up then theoretically the length of the pendulum would remain unchanged. In practice, the simplest way in which this can be achieved, and indeed that described by Harrison, is by employing several shorter sections, but keeping the ratio the same. Thus the five, seven or nine rod gridiron pendulum was evolved, of which the last was most suited to the relative expansions of brass and steel.

The gridiron became the most popular form of compensated pendulum for use in observatory regulators made in England during the eighteenth century, but by the following century the mercurial pendulum had gained the ascendancy. Although the mercurial pendulum is easier to construct and also to vary its degree of compensation by adding or removing a little mercury, and has certain other advantages, it is far more difficult to transport. It is the latter which weighed against it, particularly as regulators were being exported to many overseas countries and were used on scientific expeditions such as the Transits of Venus.

In France the gridiron pendulum, which was usually made on a far more massive scale than in England to suit the superb cases in which they were housed, was preferred to the mercurial. One suspects that this may have been, at least in part, so that the clockmakers could impress their wealthy clients who in turn could impress their friends.

John Ellicott's Compensated Pendulum

Ellicott, like Graham, took a keen interest in compensated pendulums and presented his research to the Royal Society in 1736.[3] He found the coefficients of the different metals as follows:

Gold	Silver	Brass	Copper	Iron	Steel	Lead
73	103	95	89	60	56	149

However, he realized that these figures varied appreciably with tempering, hammering, polishing, etc. and thus it was necessary to measure the coefficients after they had been prepared. As a result of this research he devised two different forms of compensated pendulum, his descriptions of these being contained in Chapter 12.

The advantages of Ellicott's pendulum over the gridiron are that the degree of compensation may be varied, and that the bob is supported at its mid-point and thus any expansion or contraction by it would be self-canceling. It was also thought to overcome the problem that often occurs with the gridiron of the rods sticking in the blocks they pass through; however Ellicott's own designs of compensated pendulum were also said to suffer from this problem.

During the next one hundred fifty years numerous different forms of compensated pendulum were to be devised, some of which will be discussed later.

As the accuracy of the thermal compensation increased towards the end of the nineteenth century, one other factor came into play. The effect of stratified temperature, i.e. the changing temperature at different levels in a room due to hot air rising as exemplified by the air at ceiling level being appreciably hotter than that near the floor. This means, in effect, that the temperature will also vary along the pendulum rod and ideally should be allowed for. This will be discussed at a later stage.

Barometric Pressure and Humidity

Although it had been realized at a comparatively early period that fluctuations in barometric pressure affected timekeeping, it was not until the second half of the nineteenth century that serious efforts were made to compensate for this by incorporating barometric compensation units such as that used by Dent (Fig. 4-2) and at a later date by Riefler (Fig. 4-4); however at a much earlier period Harrison had already taken this into consideration when designing his gridiron pendulum.

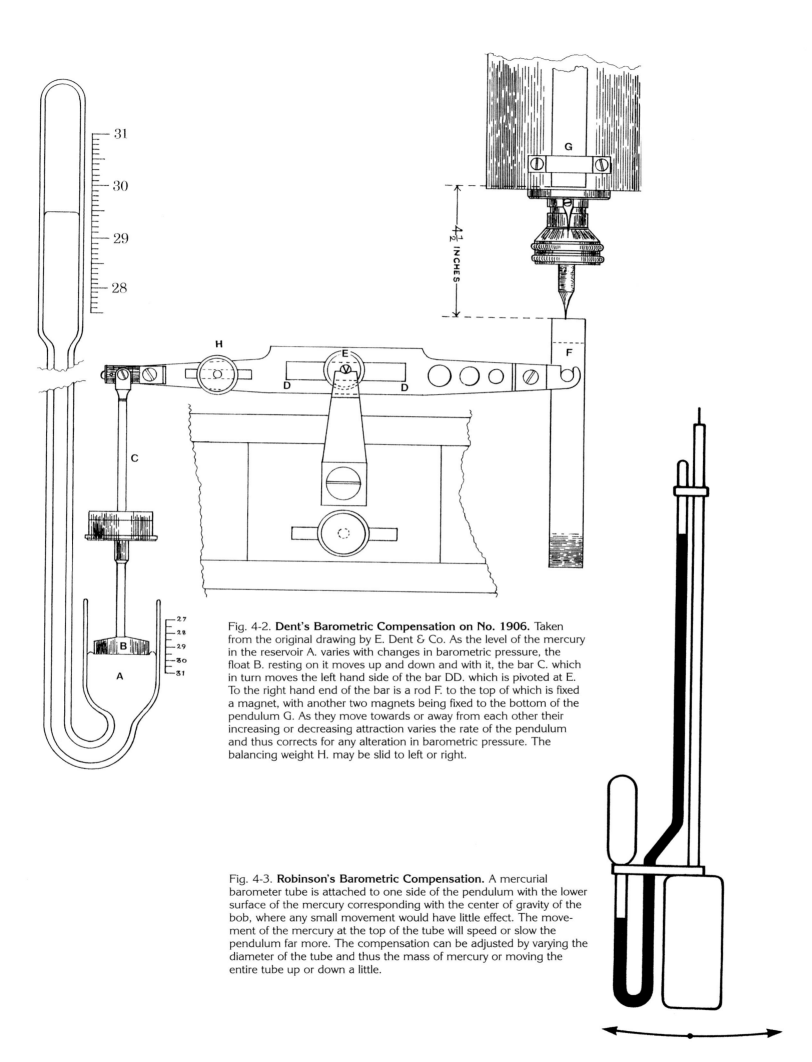

Fig. 4-2. **Dent's Barometric Compensation on No. 1906.** Taken from the original drawing by E. Dent & Co. As the level of the mercury in the reservoir A. varies with changes in barometric pressure, the float B. resting on it moves up and down and with it, the bar C. which in turn moves the left hand side of the bar DD. which is pivoted at E. To the right hand end of the bar is a rod F. to the top of which is fixed a magnet, with another two magnets being fixed to the bottom of the pendulum G. As they move towards or away from each other their increasing or decreasing attraction varies the rate of the pendulum and thus corrects for any alteration in barometric pressure. The balancing weight H. may be slid to left or right.

Fig. 4-3. **Robinson's Barometric Compensation.** A mercurial barometer tube is attached to one side of the pendulum with the lower surface of the mercury corresponding with the center of gravity of the bob, where any small movement would have little effect. The movement of the mercury at the top of the tube will speed or slow the pendulum far more. The compensation can be adjusted by varying the diameter of the tube and thus the mass of mercury or moving the entire tube up or down a little.

Fig. 4-4. **Riefler's Barometric Compensation Unit.** This is mounted near the top of the pendulum rod. As it expands up with decreasing pressure, it carries with it the compensating weight resting on top. The scale to the left indicates the degree of compensation being achieved.

There are three main effects of rises in barometric pressure: (a) the "flotation factor", i.e. if the barometric pressure rises the air gets denser and thus the effective weight of the pendulum is reduced, which will slow it; (b) there is an increase in the drag on the pendulum due to the air which is moving too and fro with it, which again will reduce the rate; (c) the increase in friction of the air also reduces the rate, but only by a very small amount.

The effect of the above factors will vary with the shape and density of the pendulum bob. The greater its density the less the variation with rises in barometric pressure; however, it is unlikely that the above factors will cause a deviation of more than 0.35 seconds a day, and this will be offset to some degree by the reduction in arc of the pendulum and a corresponding change in the circular error, which was Harrison's technique.

In theory it should be possible to design a pendulum in which all these factors cancel each other out, thus eliminating any errors in timekeeping due to fluctuations in barometric pressure, but whether this is achievable is open to debate[4], although Harrison is thought by some to have done so.[5]

The two other factors to be borne in mind are that a rise in temperature will decrease the density of the air and a rise in humidity will have the same result; however the maximum effect of the latter is unlikely to exceed 0.18 seconds a day. The effects of changes in barometric pressure may be of a short term nature, for instance a passing storm, or, on an annual basis, with barometric pressure for instance, being higher in winter than summer. The effects of this have been analyzed by J. E. Bigelow.[6]

Correcting for Barometric Pressure

The obvious way of eliminating the effect of changes in barometric pressure is to enclose the pendulum within a sealed container maintained at a constant pressure. With the Le Roy and Riefler tank regulators, the pressure is maintained at 25-26" mercury, whilst with the Shortt, it is reduced to about 1", which would seem the better solution as a low barometric pressure increases the accuracy, but why this should be so is not completely clear. At the same time temperature compensation can be virtually eliminated by maintaining a constant temperature within the jar. If this is not done, then not only will the pendulum be directly affected by the changing temperature, but also indirectly because any temperature fluctuation will also affect the density of the remaining air in the cylinder.

The running of a clock in a vacuum had been mentioned as early as the seventeenth century by Matteo Campani (1620-1687) and in 1704 William Derham had conducted experiments running a clock in a vacuum and found that the arc of vibration increased in proportion to the reduction of air pressure; however the difficulty was that at that time no method could be devised to wind it; a problem that was only resolved in the mid-nineteenth century with the advent of electricity.

An alternative method of barometric compensation was devised for Dent No. 1906 which was made, with Airy's co-operation, for the Royal Observatory in Greenwich c. 1870. This clock, which was fitted with Airy's escapement, proved to be so accurate that it actually recorded changes in barometric pressure.

To eliminate this error Dent devised an ingenious system whereby a float was placed on the surface of the mercury of an ordinary barometer (Fig. 4-2). Attached to this is a pivoted lever on the other end of which is a magnet, placed immediately below the pendulum, to the bottom of which two further magnets were attached. As the barometric pressure rose and fell so the magnets were moved closer together or further apart, thus varying the rate of the pendulum. This was found to be very successful.

A different method had been used by Robinson of Armagh some forty years earlier.[7] He attached a mercurial barometer tube (Fig. 4-3) to the side of the pendulum in which the lower level of the surface of the mercury corresponded with the centre of gravity of the bob, and thus a small movement of that surface would have little effect on the rate, which would be far more affected by the movement of the mercury at the top of the tube. The bore of the tube would also affect the amount of mercury being raised or lowered and this could be varied, once the mass of the pendulum was known, to achieve the correct compensation. Fine tuning was carried out by raising or lowering the tube a little.

The method adopted by Riefler, which was probably the most accurate, employs a vacuum tank with a carefully calculated weight on top of it which rises and falls in inverse proportion to the barometric pressure (Fig. 4-4).

Gravity

The gravitational force exerted on any point on the surface of the Earth is directly related to the distance of that point from the center of Earth. Because, as Newton predicted when he published his Theory of Universal Gravitation to the Royal Society in 1686, the Earth is an oblate spheroid, i.e. somewhat flattened at the poles, it means that gravity will be at a maximum at these points where it is some thirteen and one half miles closer to the center of the earth than at the equator.

The decrease in gravity at the Equator will result in a pendulum swinging more slowly there and this is magnified by the centrifugal force at this point which will account for 2/3rds of the total change. This is scarcely surprising when one bears in mind that the Earth's surface at the equator is traveling at 1,080 miles per hour.

Various other related forces will also affect the pendulum but they are of a relatively minor, but complex, nature and will not be considered here.

Probably the first person to record the effects of variations in gravity was Jean Richter, a French astronomer, who visited Cayenne near the equator. On arriving there he set up his clock, which had kept good time in France, only to find that it was losing time and had to be regulated. When he returned to France he found the reverse was true and it was gaining. These results he attributed to the Earth's rotation.

In 1732 Graham decided to carry out a series of experiments in Jamaica to assess the effect of gravity at different places and these are referred to and discussed by Charles Aked [8], his article being based on correspondence between George Graham and James Bradley. The upshot of these experiments was that Graham calculated that the pendulum went two minutes and six seconds slower in Jamaica than London, which was to go a long way to confirming Newton's theory, although he was, in fact, over-estimating the difference by forty-two seconds. This was probably in large measure due to the fact that he was using an uncompensated pendulum, probably for ease of transport, and had tried to calculate the changes that a difference in temperature would have made.

The extreme in variations produced by a pendulum swinging at the North Pole and the Equator, gives a time difference per day of three minutes and forty-six seconds, which equates to a change in length of the pendulum of approximately 0.12". To try to estimate the force of gravity by measuring the length of the pendulum is virtually impossible, bearing in mind that a change of one second a day is brought about by a change in length of just .001". Moreover, if a steel strip suspension is used, the exact point at which it flexes cannot easily be determined. Thus the only accurate way of measuring the force of gravity is by knowing the number of swings the pendulum will make in a set period, usually a day.

Those who would like to read a more detailed analysis of the effect of the Earth's rotation on a pendulum are referred to the article of Stuart Kelley.[9]

The investigations in the variations in timekeeping due to gravity were continued for the next one hundred or more years. For instance, two regulators, one sidereal and the other showing mean time which kept good time at Edinburgh were moved, without any alteration, to London, when one lost eighteen seconds and the other twenty-one seconds. Another example is that of Captain Phipps who, on his voyage towards the North Pole in 1772, tested a pendulum by Graham which had been set up to vibrate seconds in London. He set it up on an island at latitude 79° 50' north, when it was found to accelerate by seventy-two or seventy-three seconds a day.

Rees',[10] that fount of knowledge, goes into the matter in considerable detail quoting from the work of Professor Bridge of the East India College, contained in his "Introduction to the Study of the Mathematical Principles of Natural Philosophy 1813" and explains how to calculate the changes with latitude, the table he includes being given in Fig. 4-5.

Degrees of Latitude.	Length of a Pendulum.	Length of a Degree.
	Inches.	Miles.
0	39.027	68.723
5	39.029	68.730
10	39.032	68.750
15	39.036	68.783
20	39.044	68.830
25	39.057	68.882
30	39.070	68.950
35	39.084	69.020
40	39.097	69.097
45	39.111	69.176
50	39.126	69.256
55	39.142	69.330
60	39.158	69.401
65	39.168	69.467
70	39.177	69.522
75	39.185	69.568
80	39.191	69.601
85	39.195	69.620
90	39.197	69.628

Fig. 4-5. **Emerson's Table Reproduced in Rees'**[9]. Showing the length of a pendulum beating seconds at latitudes from 0° - 90°.

As already discussed, the pendulum was accepted as the ideal way of measuring the force of gravity at any one point and was used at many different locations; however it had one disadvantage, it was affected by changes of temperature, barometric pressure and various other lesser factors. Kater (Fig. 4-6) overcame this problem in 1817 when he realized that, with a pendulum, the point of suspension and the center of oscillation are interchangeable and there is no reason why the pendulum should not swing isochronously from either end. He thus constructed a brass rod nearly four feet long with accurately made steel and agate knife edge suspension points set facing each other 39.4" apart. There was a pair of counter weights and a central adjusting screw. Thus the pendulum, adjusted by the weights, could be made to swing either way up without any difference, thus canceling out the errors. This pendulum was free, i.e. not connected to any movement, and just kept swinging by being given an occasional push, although a conventional regulator was used to monitor its swings.

Fig. 4-6. **Kater's Pendulum.** This was devised by Kater in 1817 when he realized that there is no reason why a pendulum should not swing isochronously from either end. It consists of a 4 ft. brass rod with suspension points of steel and agate at either end some 39.4" apart, together with counterweights and a central adjusting screw. By swinging the pendulum both ways up any errors would cancel each other out. It was not connected to any movement.
    It is seen here being used to measure gravity with Kater's pendulum being erected in front of that of the regulator so that when they synchronize the regulator's pendulum is completely concealed, as may be observed by the telescope on the left. The number of vibrations between the pendulums coinciding can be used to determine the force of gravity at that point.

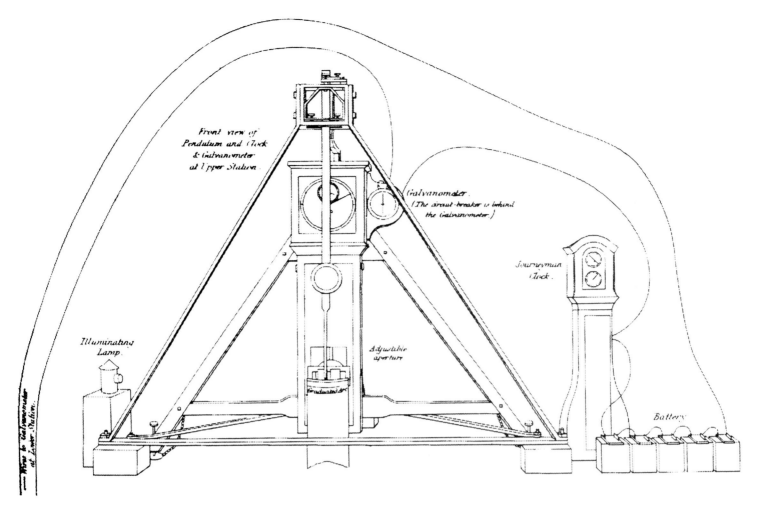

Fig. 4-7. **Airy's Experiments to Measure the Force of Gravity at the Top and Bottom of Harton Colliery in 1854.** The purpose was to measure the difference in the force of gravity at the top and bottom of a coal mine and from this determine the average density and thus the weight of the Earth. It was simplified by being able to compare the clocks at the top and bottom of the pit by means of an electric telegraph. Seen here in Airy's drawing[19] are Kater's pendulum, and the clock at the upper station and the wires leading down from it to the Journeyman clock below.

An entirely different application for Kater's pendulum was devised by G. B. Airy in 1826 (Fig. 4-7). He realized that it could also be used to measure differences in gravity on the surface of the Earth and, for instance, at the bottom of a mine where the gravitational attraction should be less because it would no longer be influenced by the mass of the Earth (i.e. its outer shell) above it. However, this assumes that the inner and outer parts of the Earth are of similar density, but it was thought that the inner part was of greater density, which would result in the pendulum swinging faster.

The main purpose of these experiments was to assess the density and thus the total mass of the Earth. The first of two experiments was carried out at Dolcoath Copper Mine in Cornwall in 1826 where they installed two Kater pendulums and regulators at the top and bottom of the mine which were interchanged at set intervals and the Kater pendulums inverted.

This experiment was brought to a premature end when one of the pendulums crashed to the bottom of the mine whilst being hoisted out. The experiments were repeated in 1828 but, due to flooding, these too were never completed.

Some twenty-six years later Airy repeated his experiments, but this time at the Harton, South Shields Coal Mine. These were far more successful, in part because of the facilities provided at the coal mine and also because the upper and lower stations were connected electrically. Within three weeks the readings had become so consistent that there was no need to continue them further.

It was then left to Airy to estimate the density of the Earth, after making allowance for any factors which might have affected the figures such as the local geology. His conclusion was that, as the depth of 1,260 feet increased the pull of gravity by 0.0000521, the density of the Earth must be 6.565 times that of water. This subsequently turned out to be an over-estimate of around 20%.

For those interested, a much fuller account of this work is given by Allen Chapman[11] and Kater reported on Airy's experiments to the Royal Society in 1818.[12]

The result of all the research into gravity and in particular its measurement throughout the world and even at different altitudes enabled clockmakers to calculate the exact length of pendulum required for their regulator to enable it to keep time wherever it was sent, Riefler being particularly meticulous in this respect.

One other use the information was put to was to assist in calculating the height of mountains. The lovely little portable regulator by Janvier (Fig. 4-8) is believed to have been designed for just this purpose.

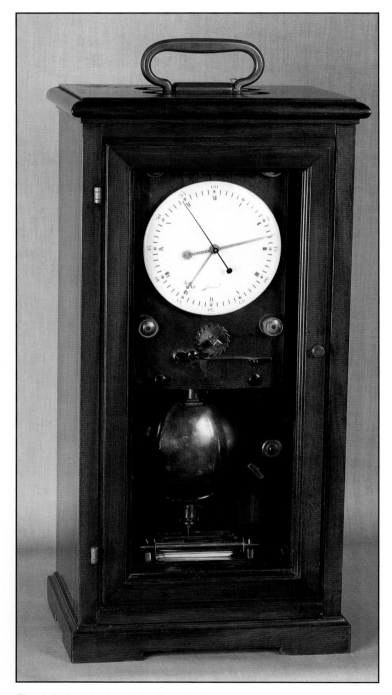

Fig. 4-8. **Antide Janvier's Portable Regulator,** believed to have been designed to assess the height of mountains.

The Gravitational Attraction of the Sun and Moon

Before leaving the subject of gravity it is as well to consider the effect of the sun and moon on a pendulum, which basically occurs in the form of tides. It has been explained in some detail by Boucheron.[13] In this he refers to the work on this subject by Loomis who in 1928 recorded the data on three Shortt clocks over a period of one year. Unfortunately, there was some inter-reaction between the pendulums, accounting for a maximum accumulated error of 0.05 seconds a week. In 1984 Boucheron carried out similar experiments, but using solid state electronics applied to Shortt No. 41 in the US Naval Observatory, Washington and presented these in the form of graphs. From these he concluded that the peak to peak amplitude change of the pendulum due to the gravitational pull of the sun and moon was of the order of 400 microseconds an hour at Washington. A cyclical variation per day of roughly +/- 1 millisecond was observed, the reason for which was unknown and there were also other fluctuations.

Humidity

This will have some effect on a pendulum as a rise in humidity will actually decrease the density of the air, although this will also be related to the temperature. The change for a rise in humidity of 50% is of the order of 0.25%.

The Suspension Spring and Knife Edge Suspension

There are two main factors which affect the pendulum's suspension spring with a rise in temperature: (a) an increase in length and (b) an increase in flexibility (modulus of elasticity). Both of these will affect the frequency of oscillation of the pendulum. The increase in length can be allowed for in the overall calculation when assessing the compensation required but the increase in flexibility is a little more difficult to assess.

Attention was paid to the suspension spring by several clockmakers over the years. Dent, carrying out some research in the 1830s, found that the stronger the suspension spring the quicker the pendulum stops and that a rise in temperature increases the elasticity of the suspension and thus the rate of the pendulum; however it also changes its length.

W. J. Frodsham also carried out research on suspension springs, delivering a paper to the Royal Society in 1838 and subsequently publishing his results privately. He used varying lengths and thicknesses of suspension together with varying weights of maintaining power. He showed that if you used the correct suspension, the pendulum would stay isochronous, even with a change of weight.

The ideal solution came in the late 1930s when Fedchenko designed and produced the first isochronous suspension.[14,15]

It might be thought that the ideal suspension would be the knife edge or even point suspension, but there are snags!

If point suspension, and to a lesser extent knife edge suspension, is used then its length has to be taken into account and allowed for in any calculation of temperature compensation but, possibly of more importance, wear will occur which will alter the effective length of the pendulum and contamination and corrosion may well take place. This is particularly likely to occur if iron on iron or steel on steel is used, especially if their chemical compositions are different. Ideally, agate or a similar material should be employed.

The Pendulum's Support

From our early days of learning we were taught that action and reaction are equal and opposite and so it is with the pendulum. The greater the mass of the pendulum relative to its support, the more influence it will have on it and in return the more the pendulum will be disturbed. The weight of the pendulum and in particular the bob cannot be reduced, otherwise it is more likely to be affected by outside factors and thus, for accurate timekeeping, the mounting must be as rigid and massive as practical. With the domestic regulator, the best that can be achieved is to move the mounting bracket to the backboard and then bolt this to the wall. Dent understood the problem well and provided a very heavy cast iron bracket to carry the pendulum and in an observatory situation sometimes bolted this to the wall, or a stone slab, leaving the case freestanding around the movement and pendulum (Fig. 4-9). He also often made his cases flat bottomed so that they could rest on a stone corbel.

Fig. 4-9. **A regulator attributed to Dent**. In which the movement and pendulum are supported by a massive cast iron bracket bolted directly to the wall for maximum stability with the case standing around it but not attached in any way.

Professor Hall's regulator, probably the most accurate pendulum clock produced, rests on a twelve ton concrete block to give maximum stability and even then its rate can be disturbed by the movement of the roots of a nearby tree during a high wind.

The Search for the Free Pendulum

The less a pendulum is disturbed the better; however it will require impulsing at regular intervals and by the same amount if it is to maintain a constant arc of swing. The difficulty here is that the impulses may vary in two ways (a) on a short term basis due, for instance, to variable depthing in the wheel train which will alter the power coming through to the pendulum and (b) over the longer term, because of wear occurring in the train and more particularly the pivot holes which will alter the depthing between the wheels and pinions and greatly decrease efficiency. A further major factor is the drying up of the oil, a particularly important factor with the lubricants obtained from animal and vegetable sources in the eighteenth and nineteenth centuries. By virtually eliminating the need for this, Harrison solved one of the major problems of accurate timekeeping.

Train Remontoires

One of the ways of mitigating these effects, which was used quite extensively in France, was to employ a train remontoire, which was first conceived by Jost Burgi some four hundred years ago. With this an auxiliary weight or occasionally a spring is wound up or tensioned at set intervals and acts on the upper part of the train to produce a relatively constant force on the escapement. One of its biggest advantages is that it can be used to convert a spring driven clock with its relatively variable power source into one that is weight driven and thus exerts a constant force. The rewind period may vary from a few seconds, to minutes or even hours.

The disadvantage of the train remontoire is that it is still vulnerable, albeit to a lesser extent, to the drying up of the oil and other factors mentioned earlier. The ideal, which overcomes these problems, is the escapement remontoire, in which a very light weight is lifted by the movement a small and constant amount and then released, usually at each swing of the pendulum, thus always providing the same impulse. Any fall off in power transmitted through the movement will have almost no effect on the impulse given until such time as it can no longer lift, for instance, the gravity arms or deflect a spring a constant amount, when the clock will stop. Although the concept is simple, the struggle to perfect it took the best part of one hundred years and is discussed in full in Chapter 6.

With the increasing application of electricity to horological problems, this was used, rather than the clock's train, to rewind the remontoire.

Mechanical Improvements

As the precision pendulum clock evolved, so the quality of the movement improved. Higher train counts were employed to reduce friction, six and eight leaf pinions being abandoned in favor of those of ten and fourteen and sometimes even higher counts, with an accompanying increase in the wheel count and their diameter. The precision with which they were cut and the use of six or more crossings were used to produce a lighter wheel, thus reducing inertial losses and providing greater rigidity.

Wear

Wear was a big factor to be considered, particularly in the eighteenth century when the quality of lubricants was poor. Apart from the actual damage caused to the movement, the performance was affected by increasing friction, causing a variation in depthing between wheels and pinions and more particularly between the pallets and escape wheel. As the efficiency of the movement decreased, so the power transmitted to the pendulum and thus its arc of swing was altered and therefore the circular error varied.

The Reduction of Friction

Clockmakers had been aware of the need to reduce friction from a comparatively early period. The reasons for this are fairly straightforward: (a) the greater the friction the greater the wear, which in turn will increase friction still further; (b) the power required, be it a mainspring or weight, will increase in direct proportion to the friction in the train and the escapement; and (c) the greater the friction, the less the power coming through the train and thus the impulse given to the pendulum and the more that power is likely to vary as, for instance, the oil dries up.

One of the difficulties early clockmakers had was the lubrication available, the only oils being animal or vegetable based, which tended to deteriorate and dry out quite quickly. Even with modern mineral oils the problem was by no means solved. Thus the only option available was to increase the efficiency of their clocks. The main way of doing this was to finish up all the wheels, pinions and pivots as meticulously as possible.

Louis Berthoud gave a lot of attention over the years to the reduction of friction, notes on this being contained in his extensive workshop books which have been reproduced and commented on by Jean-Claude Sabrier.[16] Among the areas covered by Berthoud are the manufacture of high count pinions; i.e. replacing the standard eight to ten leaf pinions with those of sixteen, eighteen and twenty. He comments on these as follows, *Evidently making them becomes infinitely longer and more difficult, but the precision of the gears and the equalizing of pressure on the arbors, amply repays the trouble.*

He also tried many different metals for the pivot holes including gold and platinum, but found little benefit, possibly in part because of impurities in the metals, and in the end resorted to jewels, preferring rubies, although he did also try diamonds. At first he imported them from England but later prepared them himself. Berthoud also used anti-friction wheels quite extensively in his watches and chronometers.

Jewelling

To try and reduce wear, jewelling was introduced to the pallets and pivot holes. This latter was most frequently employed for the escape wheel and pallet arbors where the movement was rapid relative to the rest of the train. Jewelling was also employed in the form of end stones to control the position of the arbor between the plates and prevent butting of the shoulder of the pivot against them, a factor that can cause severe friction. Other refinements that were introduced were the enclosure of the movement within either a wooden or a metal dust cover and the provision of some form of beat regulation. An indication of the efficiency of many of the regulators made in the nineteenth century is that they will run for a week, providing impulse to a pendulum weighing twelve to sixteen pounds on only a three to four pound weight; however it must be remembered that the resistance to its passage through the air is the critical factor.

Berthoud commented on the desirability of a balance (or pendulum) oscillating freely between two impulses and indeed the detaching of the pendulum from the influence of the movement and thus rendering any variation in the power being transmitted of little consequence. This was the holy grail of clockmakers for at least two hundred years.

The (Almost) Ultimate Solution and the Tank Regulator

As with so many things in life, the ultimate solution to a problem is, if possible, to avoid it. Such was to be the case with the precision pendulum clock. Once a system of rewinding a clock within a sealed tank had been devised, a relatively simple matter with the aid of electricity, several problems were solved. The first was that the clock could be kept at a constant barometric pressure, thus protecting the pendulum from a variable resistance to its passage and overcoming the effects of changing "flotation." The second problem overcome was that with the aid of a thermostat, the air within the tank can be kept, thanks to Harrison and his bi-metallic strip, at a constant temperature, thus overcoming the problems of any changes in the length of the pendulum with temperature and, less importantly, the variations in the density of the air caused by this. The only problems remaining so far as temperature is concerned are those of stratification, i.e. different temperatures at different levels, which are less likely to occur in a tank than a room; however Riefler largely overcame this problem in the early twentieth century.

A further advantage of constant temperature is that the suspension spring will stay the same length and flex in the same manner.

The only relatively minor problems remaining when enclosing a regulator in a tank with constant temperature and barometric pressure were:

a. That the pendulum is not entirely free, i.e. it still had to unlock the escapement, until such time as Shortt developed his concept of the slave pendulum which has to do all the work and is brought back into synchronisation with the master pendulum from time to time.

b. Metals do not slide together well in a vacuum as their protective oxide layer, once lost, cannot reform and thus the parts rubbing on each other will tend to weld together by crystal growth.[17] The only way of avoiding this is to use widely dissimilar metals.

c. Any oil used will tend to evaporate in a vacuum. Both (b) and (c) will only apply if the pressure is reduced to a low level.

d. Invar, which has been used for the pendulum rods of most of the finest observatory regulators during the twentieth century because of its very low coefficient of expansion, is not entirely stable. Even after annealing it will continue to slowly change in length, usually in small jerks, for many years, albeit at a gradually reducing rate. Quartz is far better in this respect but more difficult to handle.

Maintaining Power

As the precision of clocks improved so the time lost during winding, when the power was removed from the train, became increasingly significant. A further, although relatively minor matter at the time, was the decrease in arc of the pendulum because it was receiving no impulse and would take some time to settle back to a constant rate. Another factor was possible damage to the escapement, and in particular, the relatively delicate and sometimes jewelled pallets used on regulators, if train reversal occurred. During winding without maintaining power, the escape wheel will be stationary and the pallets still moving to and fro, carried by the pendulum, may butt into the escape wheel teeth, which is why maintaining power is used, even on some year duration clocks.

A form of maintaining power was being used on turret clocks from an early date but this was not suitable for longcase clocks or regulators. Various different designs were employed but the basic principle consisted of a pivoted lever with a weight at one end to provide the driving force and a nib at the other which is moved into engagement with one of the wheels in the train, usually the third, prior to winding. As the wheel rotates it gradually comes out of mesh with the nib. (Fig. 4-10).

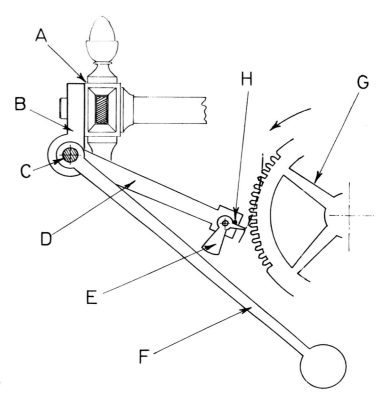

Fig. 4-10. **Turret Clock Maintaining Power.** The drawing, by John Martin, shows the maintaining power arrangement fitted to a Charles Frodsham clock, dated 1854 and views the movement from the rear.

A is the top corner and finial of the front frame, and, bolted to the outside of that corner post is a bracket **B** which carries an arbor **C** which spans the frames and projects from the rear of the movement being carried in a similar bushed bracket to **B** on the rear corner post. The arbor carries an arm **D** mounted inside the rear frame. This arm has, freely pivoted on it, a weighted latch **E**. A further arm **F** is fitted to the arbor outside the frame and this is of cast iron with an integral flat disc weight.

In order to engage the maintaining power, the arm **F** is lifted by hand and the nib of the latch **E** clicks over several teeth of the second train wheel **G**. On releasing **F** the nib bears down on a tooth and is restrained from turning anticlockwise by a pin **H**. The weight of the whole assembly provides the power necessary to allow winding of the train. The nib finally disengages and the arbor carries a stop piece to allow the arms to remain in the position shown.

Huygens' Endless Rope

This was invented by Huygens c.1658 and became a standard feature of English thirty hour longcase and lanterns clocks. It was also used on a few regulators, both in this country and in France. (Fig. 4-11)

Bolt and Shutter Maintaining Power

This particular form of maintaining power was employed extensively on longcase clocks until around 1700 after which it gradually fell out of favor. This is probably because of its unreliability which was seldom due to the maintaining power itself but more often to the shutters which were easy to damage or bend whilst inserting the key when winding the clock and often got stuck. However some top makers such as Graham continued to employ it until the 1750's and it went on being used in regulators until the end of the century. Once Harrison's maintaining power, known as the going ratchet, started to be fitted to regulators, bolt and shutter was seldom used.

With this form of maintaining power, a cord is pulled or a lever depressed which actuates a spring loaded bolt which engages on a wheel in the train, usually the third wheel and continues to supply power when the force of the driving weight is removed during winding. To make certain that this is always employed, shutters are used to obscure the winding holes and these are automatically removed when the maintaining power is actuated, thus permitting the clock to be wound. (Fig. 4-12A, B)

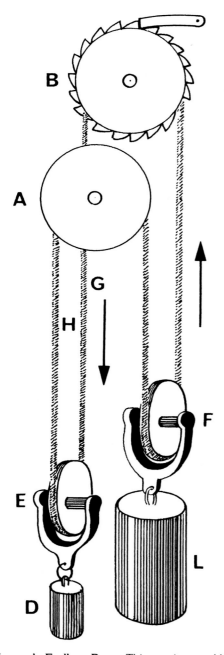

Fig. 4-11. **Huygen's Endless Rope.** This was invented by Huygens c. 1658 and came into general use on 30 hour wall, longcase and lantern clocks. The design is such that power to the train is always maintained, even during winding.

From H the endless or continuous rope passes over the pulley A which is attached to the great wheel and then passes down to and round the pulley F carrying the weight L, and up and over the ratchet wheel B. It then goes down and under the pulley E carrying the counterweight D and from there returns to the pulley A. The pulleys A and B have steel spikes to prevent the rope from slipping. Pulling down on the cord at G will raise the weight L which will still continue to exert force on the pulley A and keep the clock going.

Fig. 4-12A, B. **Bolt and Shutter Maintaining Power.** With this the clock cannot be wound until a lever is depressed or a cord pulled which will remove the shutters from in front of the winding holes and at the same time actuate a spring loaded bolt which engages on the tooth of a wheel in the train and continues to supply power when the force of the driving weight is removed during winding.

Harrison's Maintaining Power

The disadvantage of "bolt and shutter" was the inconvenience of having to actuate it by pulling a cord prior to winding and from Harrison's point of view when designing his sea clocks, it had the serious disadvantage that if it received a sudden jolt, as could easily happen at sea, then the bolt might well become disengaged from the wheel it was driving. To overcome this, Harrison employed an additional wheel with fine teeth and its own click next to the great wheel (Fig. 4-13A, B). This was linked to the great wheel by a spring or springs which, when the power was removed from the train, still applied pressure to it and thus kept the clock going.

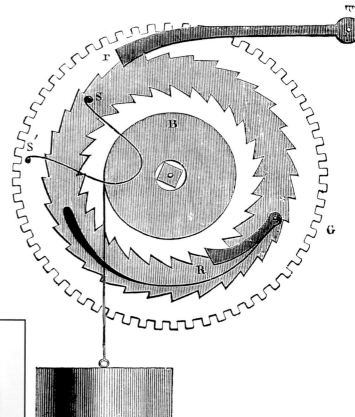

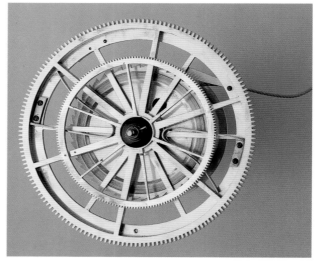

Fig. 4-13A, B, C. **Harrison's Maintaining Power. (Drawing taken from Beckett's 'Clocks, Watches & Bells').** This was devised by Harrison for his sea clocks, but eventually came into general use on regulators. Its big advantage over bolt and shutter is that it operates automatically. The ratchet wheel r, which has a click R attached to it, is connected to the great wheel by the spring $s - s^1$. Whilst the clock is running the weight acts on the great wheel and through this the spring, but when the clock is being wound and the force of the weight is removed, the click at the top Tr, which is pivoted on the frame, prevents the ratchet wheel from falling back and thus it still drives the great wheel.

In the side view of the movement (Fig. 4-13B) the additional wheel, a little smaller than the great wheel, is for the maintaining power with its detent above it. In Fig. 4-13C is Harrison's maintaining power as seen on a regulator by Leyland which is illustrated in Fig. 18-13.

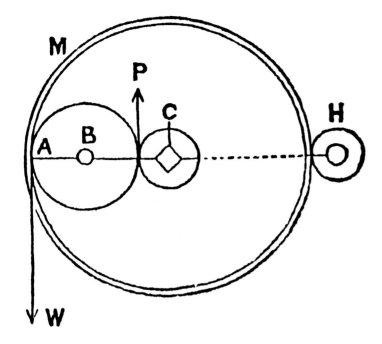

The power of the spring is gained by its compression by the mainspring or weight and on some regulators, this ideally should extend over two or three teeth of the maintaining wheel. The finer the teeth on this, the more that can be encompassed and thus, the longer the power which is available.

Harrison's form of maintaining power superseded the "bolt and shutter" and was probably employed on at least 90% of all regulators made after 1790.

Epicyclic Maintaining Power

It was in 1683 that Huygens first described epicyclic maintaining power, (epicyclic gearing had been in use much earlier than this) which is automatically provided if epicyclic winding is used. However, it never gained general acceptance, in part because of the difficulty and expense of manufacture and also because of the limited power transmitted, although this was probably adequate for most regulators. According to Beckett[18] (Denison) it was unsuitable for turret clock work because of its short duration and unless the gears were very well made they had a tendency to jam; however it was used by Arnold for chronometers.

Its use continued in limited numbers right up until the twentieth century. Cooke of London and York, for instance, the famous scientific instrument makers, incorporated it in some of their regulators. (Fig. 4-14A, B).

Figs. 4-14A, B. **Epicyclic Maintaining Power.** This never came into general use but was employed by some regulator makers such as Cooke of York and Moore of Leeds. In the Cooke regulator seen here, pins have been used for the annular ring and the teeth of the sun and planetary wheels may just be seen extending beyond the wide crossings.

In the drawing, taken from Beckett's 'Clocks, Watches & Bells', a rim at the back of the great wheel M has internal teeth cut in it, which engage at A on the wheel pivoted at B which meshes with the pinion C. The arbor of this runs loose through the barrel ends. When the movement is wound some force is applied to the intermediate wheel and this is the force necessary to lift the weight.

Other forms of maintaining power were designed over the centuries but none had the relative simplicity or reliability of Harrison's and some had to be set, as with "bolt and shutter" each time the clock was wound. An example is Thacker's clock in which a separate winding square was provided for the maintaining power. Another form of bolt and shutter power seen on some French clocks in the nineteenth century is the use of the force applied when inserting the winding key to tension an auxiliary spring.

References

[1] Robert, M.H. *The retardation of a pendulum caused by an increase in its arc of oscillation*. Taken from 'Treatise on Modern Horology' by Saunier. p. 692. Reprinted 1975.

[2] Graham, G. *A contrivance to avoid irregularities in a clock's motion, occasioned by the action of heat and cold upon the rod of the pendulum*. Philosophical Transactions. 34. 1726. pp. 40-44.

[3] Ellicott, J. *The Description and manner of using an instrument for measuring the Degrees of the Expansion of Metals by Heat*. Philosophical Transactions. Vol. 39 pp. 297 -299. Abridgements 1809 (submitted to Royal Society October 1736).

[4] Robertson, D. *The Theory of Pendulums and Escapements*. Horological Journal, March 1929. pp. 194 - 196.

[5] Burgess, M. *The Scandalous Neglect of Harrison's Regulator Science*. pp. 262 - 263, contained in *The Quest for Longitude*. Ed. W.J.H. Andrewes. Harvard University, 1993.

[6] Bigelow, J. E. *Barometric Pressure Changes and Pendulum Clock Error*. Horological Journal', August 1992. pp. 62 - 64.

[7] Robertson, D. *The Theory of Pendulums and Escapements*. Horological Journal, April 1929. p. 215.

[8] Aked, C. *Three letters from George Graham*. Antiquarian Horology, Winter 1988. pp. 597 - 605.

[9] Kelley, C. S. *Effect of the Earth's Rotation on the Motion of the Simple Pendulum*. Horological Science Newsletter, April 1996.

[10] Rees, A. *Rees's Clocks, Watches and Chronometers (1819-20)*. A selection from The Cyclopaedia or Universal Dictionary of Arts, Sciences and Literature. pp. 231 - 237. Reprinted by David and Charles, 1970.

[11] Chapman, A. *The Pit and the Pendulum: G. B. Airy and the Determination of Gravity*. Antiquarian Horology, Autumn 1993. pp. 70 - 78.

[12] Kater, H. *An Account of Experiments for Determining the Length of the Pendulum Vibrating Seconds in the Latitude of London*. Philosophical Transactions, 1818. p. 102.

[13] Boucheron, P. H. *Effects of the Gravitational Attractions of the Sun and Moon on the Period of a Pendulum*. Antiquarian Horology, March 1986. pp. 53 - 63.

[14] Woodward. P. *Fedchenko's Isochronous Pendulum 1*. Horological Journal, March 1999. pp. 82 -84.

[15] Woodward, P. *Fedchenko's Isochronous Pendulum 2*. Horological Journal, April 1999. pp. 122 - 124.

[16] Sabrier, J.C. *La Longitude en Mer à L'heure de Louis Berthoud et Henri Motel*' Antiquorum, Genève, 1993. pp. 287 - 294.

[17] Bigelow, N. *Friction in Horology & Elsewhere*. Horological Science Newsletter, April 1993. p. 2.

[18] Beckett, E. *Clocks, Watches and Bells, 7th Edition 1883*,. Crosby, Lockwood & Co. pp. 151, 152.

[19] Airy, G.B. *Account of Pendulum Experiments undertaken at Harton Colliery, for the purpose of determining the Mean Density of the Earth*. Phil. Trans. Royal Society 146, pp. 297-342

Chapter 5
Compensated Pendulums
by Derek Roberts

Mercurial Compensation

George Graham commenced, c. 1715, to measure the coefficients of expansion of the various metals that might be used in the construction of a pendulum so that they automatically compensated for any temperature changes. He had hoped that if one had a much higher coefficient than the other, maybe five to eight times higher, then a comparatively short length of this could be used as the bob and its much greater expansion up from the regulation nut it rested on would compensate for the increase in length of the rod. Unfortunately he found that there was relatively little difference in the rates of expansion of the various metals tried and thus he shelved this line of investigation.

In 1721 he resumed his trials testing various other materials which might be used as a pendulum bob, including mercury. He had thought that the latter, because of its high density, would have a low coefficient of expansion and was thus surprised, when he placed it near a fire, to find that it was very high with a thermal coefficient roughly six times as great as steel. He immediately constructed a pendulum incorporating the mercury in a glass jar as the bob and tested it in a clock which he placed in a room of his house which was most exposed to changes of temperature, being without a fire in the winter and because it was south facing, most exposed to the sun in summer.

He tested his clock (Fig. 5-1) against another of fine quality with a sixty pound pendulum which vibrated only 1-1/2 degrees. This had, in another more favorable situation in the house, never altered by more than fourteen seconds in twenty-four hours. However, in the test room, where temperatures between summer and winter changed more dramatically, the timekeeping fluctuated by up to thirty seconds.

Fig. 5-1. **Graham's Mercury Compensated Pendulum c. 1727** that is shown here on his regulator in the British Museum.

The trial commenced on 18th December 1721, but by January 3rd he realized his new pendulum was overcompensating and thus by the 8th January had fitted it with a shorter glass jar. This proved to be a little too short and on the 9th June a slightly longer one was fitted. At the same time, encouraged by the results being achieved, he finished up the pendulum more carefully.

From this time until 14th October 1725, over three years later, the clocks were kept going continuously without any adjustments to the hands or pendulum. Their timekeeping was checked, as and when weather permitted, against the fixed stars, which Graham reckoned gave an error of no more than two seconds. Each day he recorded the temperature and the difference in time between the two clocks; the error of the one with mercurial pendulum only varying between 1/6th and 1/9th of that of the other clock, even though the column of mercury was still somewhat short and thus undercompensating a little.

Interestingly in July 1723 Graham removed the pendulum from the heavy standard clock and replaced this with one having mercurial compensation, but using a brass jar, lacquered inside to protect it from the mercury, instead of glass. It was to be over one hundred years before Dent introduced a metal (cast iron) jar for mercury pendulums because of the much better heat transference achieved than with glass.

Graham's summarized conclusions are worth noting here: *When the pendulum is well made and is very heavy and has a narrow arc of swing there will be little inequality in its motion besides that caused by heat and cold. By using quicksilver for the pendulum and varying the diameter of the vessel which contains it or the thickness of the pendulum rod this problem should be overcome. Care must be taken to make certain that air bubbles are eliminated from the mercury when filling otherwise their sudden and great expansion may effect the timekeeping.*

This form of thermal compensation has the advantage that it can be finely regulated by the addition or removal of small quantities of mercury in the jar. A disadvantage, particularly of the earlier mercurial pendulums, is the use of glass to contain the mercury, since this is a very poor thermal conductor. Moreover, since the temperature differential is usually relatively small, it takes a considerable amount of time for changes to be transmitted to the mercury and thus for compensation to occur. It seems likely that thermal compensation achieved over a short period, for instance a hot summer day with a variation between midday and midnight of 25 degrees Fahrenheit, would be particularly poor, which could be relevant as timekeeping on a day-to-day basis is important in astronomy and, for instance, shipping. However, it is likely that over a longer period the errors produced by poor thermal conductivity would tend to cancel themselves out and certainly the compensation for the changes in temperature throughout the seasons may well be relatively satisfactory.

Interestingly in Rees[1] reference is made to the jar designed by Mr. Rows of Islington for the mercurial pendulum fitted to the regulators that Hardy supplied to the Greenwich Observatory. The details are as follows:

This cylindrical vessel was 7.5" long x 2" in diameter excluding the thickness of the glass which was .125". The height of the column of mercury suggested by Hardy was 6.4375" but after trials at the Observatory it was found necessary to increase this by 1.0625". Following this the maximum deviation recorded on any two days between summer and winter was 0.4 sec.

To improve the performance of the mercurial pendulum by speeding its response to thermal change, two variations were developed during the nineteenth century. The first was the use of two (Fig. 19-10) or more separate glass jars, thus reducing the thickness of the glass required, greatly increasing the surface area, and decreasing the mass of mercury in any one container, all factors which speed its thermal compensation. This was taken to the ultimate by Rentzsch (Fig. 5-2A, B, C). The second alternative was the substitution of cast iron or steel for glass, which greatly increases heat transfer. Interestingly in the catalogue of the 1851 Great Exhibition under Class XV Horological Instruments, the following comments appeared, together with a remarkably precise specification for the mercurial pendulum:

The compensation-pendulum now most generally employed for the best regulators, is the mercurial; and the glass jar has in most cases given way to a steel jar, the rod dipping into the mercury, so that the whole is almost simultaneously affected by changes of temperature, which cannot, however, be so with a glass jar, which is a bad conductor. This pendulum is now very generally used by the foreign manufacturers in preference to their gridiron pendulum, which hitherto had always been preferred abroad, and which they in France executed with great skill. The mercurial compensation pendulum was until lately almost unknown in France; the strict requirements, however, of science have compelled its adoption both on account of its simplicity, greater accuracy, and facility of adjustment. The following are the particulars and dimensions of the best mercurial pendulums, with steel jars, now in use:

Particulars of Mercurial Compensation Pendulums with Steel Jars, for Astronomical Clocks.

		In	‘	“
Round steel rod	{Diameter	0	3	0
	{Weight		15 oz.	
Jar	{Diameter	2	4	0
	{Inside	2	0	0
	{Length	8	7	5
	{Weight	3lb.	14oz.	
Screw threads to the inch			30	
Divisions engraved			60	
Value of a division equals about half a second daily.				
			lb.	oz.
Mercury	Weight		11	2-1/2
Pendulum	"		4	13-1/2
Total			16	0

			Inches
Suspension spring	{Length,	from	.30 to .80
	{Breadth	"	.040 to .060
	{Thickness	"	.004 to .007

Above and following page:
Figs. 5-2A, B, C. **Sigismund Rentzsch. Watchmaker to the King and Royal Family c. 1820.** This highly unusual regulator has a pendulum with a glass rod and multiple glass tubes filled with mercury resting on top of a hemispherical lead bob. The pendulum swings from front to back, not sideways and two escape wheels are employed. British Museum.

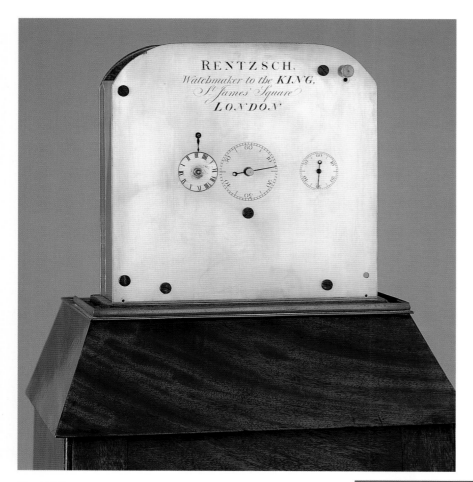

Fig. 5-2B

Fig. 5-2C

Dent and Frodsham's Pendulums

An interesting variant of the mercurial pendulum by Frodsham was described in the *Encyclopaedia Metropolitana, Vol VIII*, p. 624 in 1845, although one had been fitted to a regulator ordered by a Mr. Barlow of Woolwich for the College of Virginia, USA, which was delivered in 1826. It was different from those usually employed on two counts (a) it had an oval jar, presumably to reduce air resistance and (b) the stirrup containing the glass jar was eliminated.

This was achieved by passing an iron tube down through the centre of the jar, with the lower end being fixed securely in the bottom hole and the top end fitted into a countersink in the cover. The pendulum rod passed down through the top and bottom holes and was secured above and below by nuts which could be used to raise or lower the bob. Presumably only a few were made, as the author has never seen an example.

The first significant improvement in the mercurial pendulum after over one hundred years was that of Dent, when he adopted a cast iron (Fig. 5-3A, B) instead of a glass jar in 1836. The advantages, which he put forward in a lecture to the British Association for the Advancement of Science in 1838, was the much more rapid transference of heat, thus speeding the response of the pendulum to any changes in temperature and the far greater strength, making it much easier to transport. One further advantage was that the container could be heated to eliminate any air from the mercury.

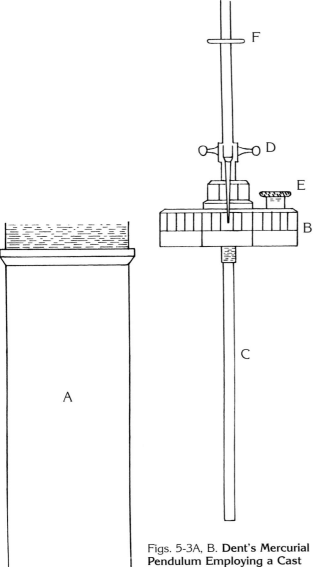

Figs. 5-3A, B. **Dent's Mercurial Pendulum Employing a Cast Iron Jar A and Screw on Cover B.** Regulation is achieved by winding the entire assembly up the rod C which passes down into the mercury to assist in temperature harmonization. The bar at D, to which a pointer is attached, is used to hold the rod and protect the suspension and the plug E may be unscrewed to adjust the level of the mercury and thus the compensation or to insert a thermometer. F is a small platform on which light regulating weights may be placed.

Charles Frodsham adopted a metal jar for his mercurial pendulums (Fig. 5-4A, B) at about the same time as Dent; which he refers to in a paper he published in 1866 titled "Mercurial Compensation Pendulums, their construction and adjustments;" however the way in which they were produced was somewhat different and more expensive, being machined out of a piece of solid steel. Their advantage was that the thin walls gave rapid heat transference, but the problems mentioned by Frodsham were that they were likely to be affected by magnetism and rust.

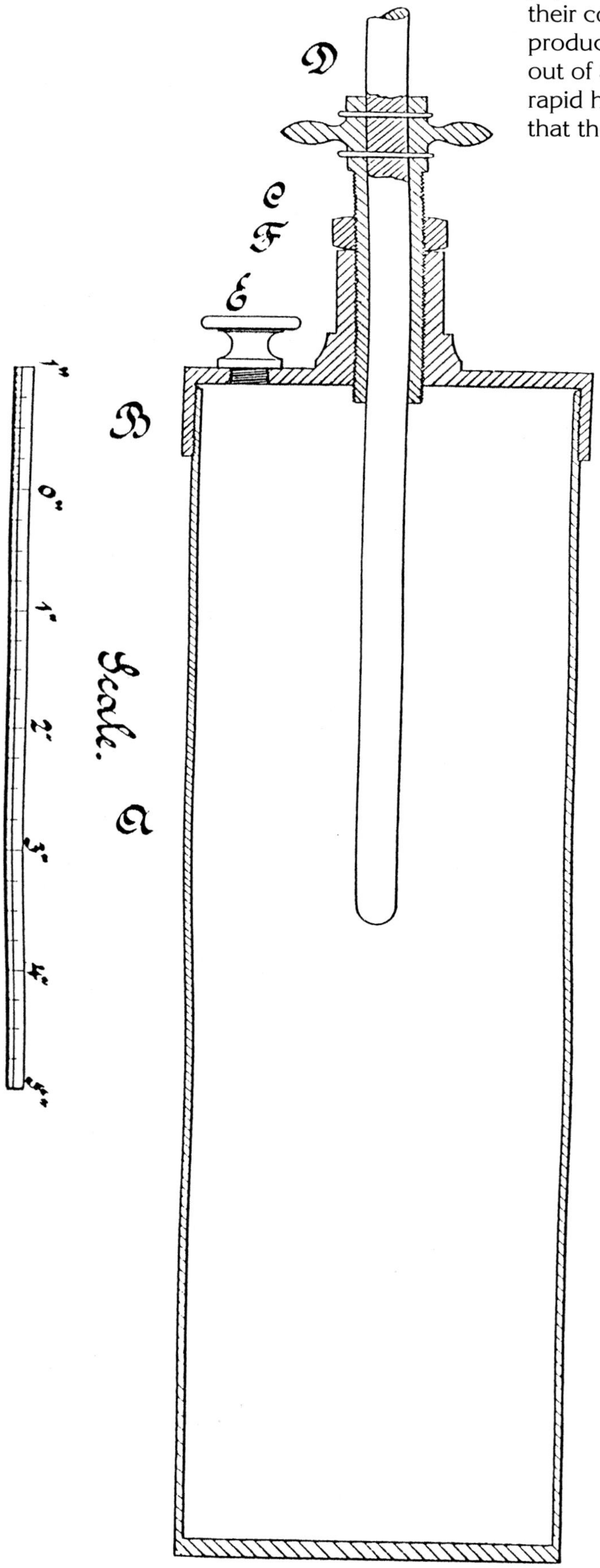

Pendulum of the Astronomical Clock, C. Frodsham. No. 1719.

Figs. 5-4A, B. Charles Frodsham's Pendulum with Steel Jar. Frodsham started using a metal jar for his mercury pendulums at about the same time as Dent but employed steel, instead of cast iron, milling it out of a solid piece, an expensive procedure, although it had the advantage that the walls were far thinner than Dent's and thus gave better heat transference. The difficulties that Frodsham mentions, apart from the high cost, were magnetism and rust and only a small number were made, their use being apparently confined to observatory regulators.

Fig. 5-4B

An alternative to the single steel jar suggested by Frodsham was a light tubular brass rod with rings attached to it into which relatively thin walled glass tubes of mercury could be slid (Fig. 19-10). Possibly the finest mercury compensated pendulum produced in England is that seen in Fig. 5-5 (and also Fig. 19-24). The use of aluminum with its relatively high rate of expansion for the pendulum rod means that the compensating columns of mercury, contained in light metal tubes, are much longer and thinner, which greatly speeds heat transference, and thus the response of the pendulum, to any changes in temperature.

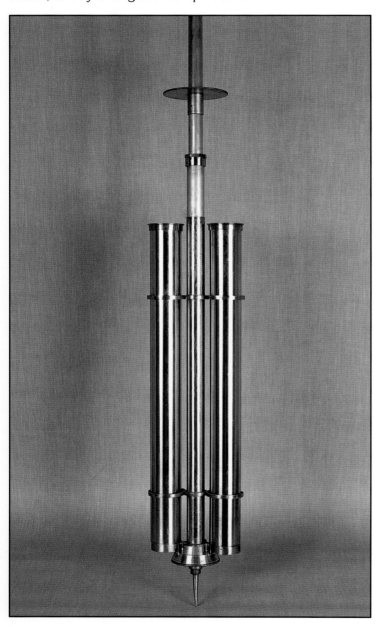

Fig. 5-5. **Frodsham's Pendulum**. This employs an aluminum rod and two relatively thin metal tubes some 11" long (see also Fig. 19-11) thus greatly speeding heat transference to or from the mercury.

The most accurate mercurial pendulum produced was that of Riefler[2] who laid down the following parameters for the design of a compensated pendulum:
1. It should have the smallest number of components and these should be able to move freely against each other with no tension or friction.
2. Temperature changes should take place simultaneously in all the components of the pendulum. Graham's mercurial pendulum is particularly poor in this respect, the rod responding rapidly and the mercury in the bob very slowly.
3. The compensating element should extend over as much of the pendulum as possible.
4. The components of the pendulum should have a high specific gravity to reduce any effect of changes in barometric pressure.
5. The pendulum should have the optimum shape for minimal resistance of its passage through the air. Riefler's mercurial pendulum (Fig. 5-6) was designed to conform with these criteria so far as possible. It was patented on the 20th March 1891.[3]

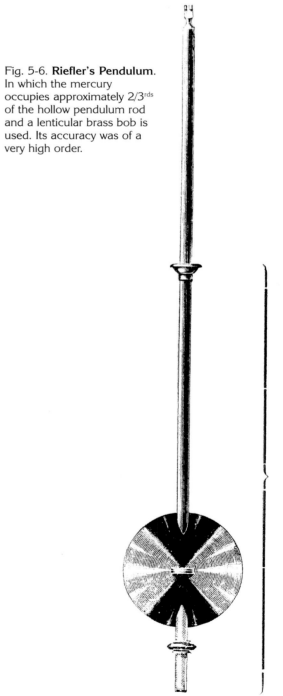

Fig. 5-6. **Riefler's Pendulum**. In which the mercury occupies approximately 2/3rds of the hollow pendulum rod and a lenticular brass bob is used. Its accuracy was of a very high order.

It consisted of a pendulum rod made of a steel tube with an overall diameter of 16-18 mm, having a 1 mm wall thickness which was filled to 2/3rds of its length with mercury, thus guaranteeing that the temperature of the rod and the mercury would stay close to each other. A brass lenticular bob was used to reduce air resistance and it had an overall weight of about 5 kg. A disc shaped weight of 110-120 grams could be screwed on below the bob that would correct the temperature compensation if the pendulum was changed from mean solar to sidereal time. It was considered important to use chemically pure mercury.

The design of the pendulum was, probably for the first time, fully calculated with the coefficients of expansion of all the components being assessed and the rods only being measured after preparation, machining and heat treatment had been carried out. Changes in air density with temperature were also taken into account.

The result of this meticulous attention to detail was that the Imperial Institute in Berlin tested the pendulum and found an error of only 0.0017 seconds per day per °C and Riefler quoted a maximum error of +/-0.005 seconds.

The production by Riefler of mercury compensated pendulums ceased on 22nd March 1902, by which time 235 had been produced. This was as a result of the discovery by Dr. Guillaume of Invar, a nickel/steel alloy with a very low coefficient of expansion.

The design, construction and testing of Riefler's mercurial pendulums is described by Dieter Riefler[2] in German and J.L. Finn and S. Riefler[4] in English.

Interestingly Riefler's concept of using a tube or column of mercury had first been used by Troughton (Fig. 5-7) some one hundred years earlier. Surprisingly it would seem that no one tried to calculate the correct measurements for Graham's mercurial pendulum, but rather determined it empirically which could prove to be a long and tedious job. Stewart[5] explains in his article how to do the necessary calculation.

Fig. 5-7. **Troughton's Mercurial Pendulum. c. 1790.** This was designed to overcome the problem of the mass of mercury in the bob and the rod responding at different speeds to changes in temperature. It consists of a strong glass tube with a bulb at the bottom similar to a thermometer, but with the bulb containing roughly 40 oz. of mercury and the tube being similar in diameter to that of a barometer; which is filled up to roughly half its length and is provided with a scale. It has the suspension fitted at the top. The bulb at the bottom is surrounded by a brass and lead bob.

The Gridiron Pendulum

By around 1725 John Harrison's work had started to change direction from making clocks of relatively standard concept, albeit with materials not normally employed in clockmaking in England, to research into their design and developing new methods of construction. Initially this was to some extent forced on him by the problems he encountered with the Brocklesby Park clock[6] and resulted in his invention of the grasshopper escapement, which required no oiling, and the employment of anti-friction rollers; however his innovations did not stop there.

It must have been around this time that his mind turned to the problem of achieving an isochronous pendulum, i.e. one which always takes exactly the same time to swing from one side to the other, no matter what the outside influences.

The most important difficulty to be overcome was the variation of the length of the pendulum with changing temperature. Graham had introduced the first practical solution to the problem with his mercury compensated pendulum a few years earlier, having discarded the use of steel and brass because their relative coefficients of expansion, roughly 3:5, were too close for one to be used in direct opposition to the other. Harrison carried out considerable research on the expansion of various metals and overcame this by using several shorter lengths of brass and steel, the combined expansions of which cancelled each other out. Thus was born the gridiron (Fig. 5-8) beneath which a conventional elliptical bob was attached. Harrison realized that a speedy response to changes in temperature was required which was one of the reasons he employed light rods. He also made the steel and brass of different thicknesses to take into account their relative density and thermal conductivity. Heavy rods such as those seen on many French regulators, whilst obviously most impressive, are a positive disadvantage so far as accuracy is concerned, not only because of their much slower response to changes in temperature but also the relatively large mass of metal being moved up and down, when this occurs.

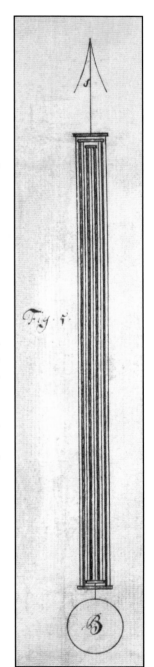

Figs. 5-8. **Harrison's original drawing of his gridiron pendulum,** taken from his 1730 manuscript. Only one example of an original Harrison gridiron pendulum exists which is shown in Fig. 10-7G and is the one on the 1728 regulator that remained with him throughout his life and is now with the Worshipful Company of Clockmakers to whom we are indebted for both pictures.

75

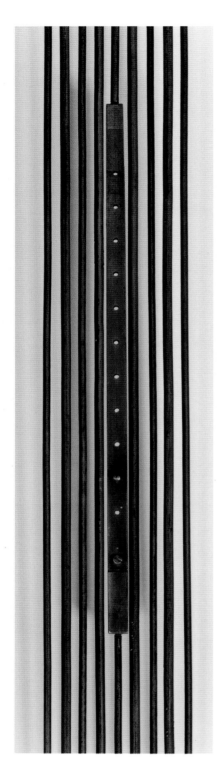

One of the criticisms of the gridiron pendulum is that the compensation may only be adjusted by dismantling the pendulum and re-pinning the rods so as to alter their effective lengths relative to each other; however Harrison overcame this problem by what Laycock[7] describes as his tin whistle (Fig. 5-9) which consists of a thin central steel bar with a brass bar on either side, all of which are pierced by a series of holes and may be locked in position, after adjusting their relative positions to correct for any inaccuracy in compensation, by inserting a pin. Martin Burgess[8], after studying Harrison's portrait in which the pendulum is included, points out that these adjustors were also incorporated in the outer rods. This method of adjusting the compensation was used by Earnshaw and one or two other English clockmakers but is seldom encountered; however it is occasionally seen in German regulators made in the following century (Fig. 5-10).

Fig. 5-9. **Harrison's device for varying the length of the central rod, and thus the compensation.** Referred to by Laycock as the Tin Whistle. It consists of a central brass bar sandwiched between two steel bars, all of which are perforated and may be locked together by a pin after their relationship has been adjusted to obtain the correct degree of compensation.

Fig. 5-10. **A regulator pendulum by Felton of Elmshorn, C. 1860.** In this the compensation may be varied by altering the effective length of the central rod.

Not only was Harrison aware of the need for compensation for variations in temperature; he also undoubtedly realized that changes in barometric pressure would affect the pendulum and it is thought that he took this into account, trying to design the pendulum in such a way that the summation of the various disturbing influences would cancel each other out. Those who are interested in this, and indeed the gridiron pendulum and Harrison's quest for accuracy on land, can do no better than read Laycock[7]; the chapters on Harrison by Andrew King[6] and Martin Burgess[8] which appeared in *The Quest for Longitude* and William Andrewes' study of Harrison's early work with a detailed account of the restoration of his unfinished clock.[9]

The first regulator produced by Harrison which employed the gridiron pendulum is believed to be the one made in 1728 which he retained throughout his lifetime and is now in the collection of the Worshipful Company of Clockmakers (Fig. 10-7). The famous claims regarding the accuracy of this regulator and the means he employed to test it are discussed in Chapter 10.

The disadvantages of the gridiron pendulum, besides the difficulty of adjusting the compensation, particularly when compared to the mercurial pendulum, are that the movement of the rods, as expansion and contraction occurs, tend to be jerky. This is because of the tendency of the rods to stick in the horizontal bars; a problem which is aggravated by corrosion with the passage of time and the ingress of protective lacquer between the block and the rod; indeed it is not unusual, when stripping down a gridiron pendulum which has not been dismantled for many years, to find that it is seized solid.

A further disadvantage of most gridiron and indeed other pendulums made in the eighteenth and nineteenth centuries is that the bob is supported from the bottom and that any expansion will take place in an upwards direction, whereas ideally it should be supported from its mid-point, when any expansion or contraction should have little or no effect; however the gridiron pendulum can be designed so that only expansion of the bob upwards is taken into account, although the difficulty there is that the mass of the bob is far greater than that of the gridiron.

Probably the two people other than Harrison who devoted most time to the gridiron pendulum were Thomas Earnshaw (see Chapter 14) and John Arnold. The early pendulums of the latter incorporated nine rods (Fig. 5-11) but those on his later examples just five (Figs. 5-12, 5-13, and 5-14). To enable him to do this he had to use zinc instead of brass rods because of their higher rate of expansion.

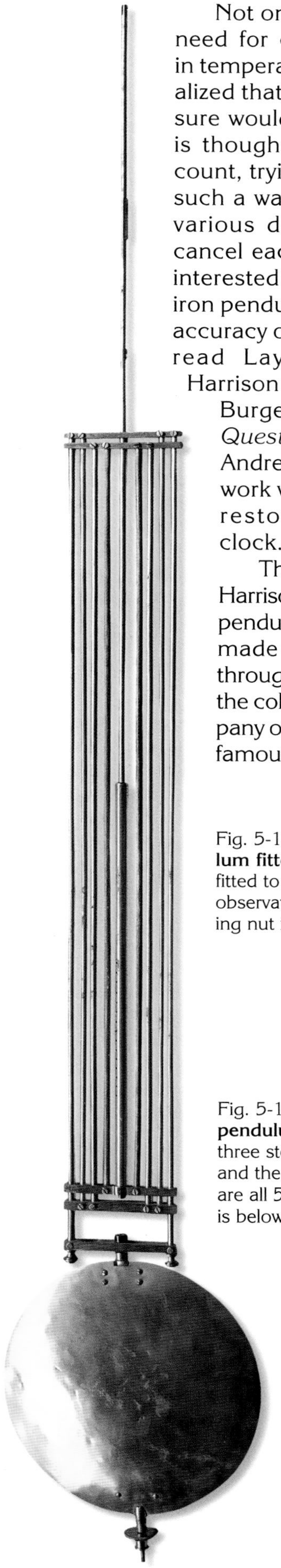

Fig. 5-11. **Nine rod gridiron pendulum fitted by John Arnold.** This was fitted to the regulator he supplied to the observatory at Budapest. The regulating nut is placed below the bob.

Fig. 5-12 **Five rod zinc/steel gridiron pendulum used on Arnold No. 34.** The three steel rods are 5.5 mm. in diameter and the two made of zinc 11 mm. They are all 550 mm. long. The regulation nut is below the bob.

Fig. 5-13. **Five rod zinc/steel compensated pendulum fitted to John Arnold No. 101.** The steel rods are 5 mm. in diameter and 710 mm. long, and those of zinc 9 mm. in diameter and 688 mm. long. The five rods extend down into the bob and the regulating nut is below it.

Fig. 5-14. **Five rod zinc/steel compensated pendulum fitted to the regulator Arnold supplied to the Observatory at Eger C. 1775.** The gridiron extends down into the bob but this is concealed by a brass cover.

Those who would like to know more about the detailed construction of the gridiron pendulum are referred to Thomas Reid[10] who gives the total effective length of the rods in the pendulum as steel 109.568" and brass 66.99".

Undoubtedly the gridiron was the compensated pendulum most favored in England during the eighteenth century for observatory regulators; even George Graham who invented the mercurial pendulum used them. Whether they were considered more accurate is open to doubt but they were undoubtedly far easier to transport than one containing mercury and in doing so their effective compensation was unlikely to be altered. These were important facts when England was the leading supplier of regulators, virtually throughout the world.

The French also used the gridiron pendulum during the eighteenth and the first part of the nineteenth centuries but its use gradually gave way to mercurial compensation and in England the majority of high grade regulators produced also used the mercurial pendulum from c. 1800.

Shown in Figs. 5-15A, B are gridiron pendulums used on French table regulators and Figs. 5-16A, B illustrates some employed on longcase regulators. It will be seen that these quite often incorporate a scale to indicate the fluctuating temperature being compensated for.

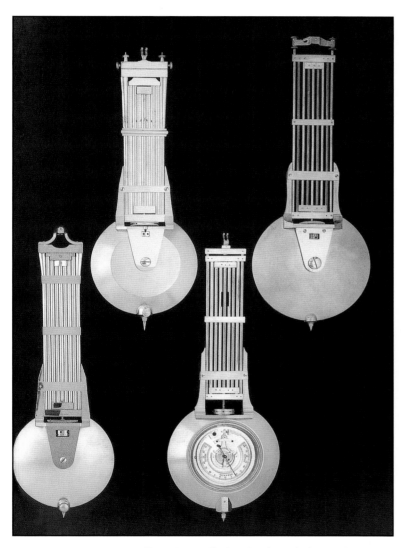
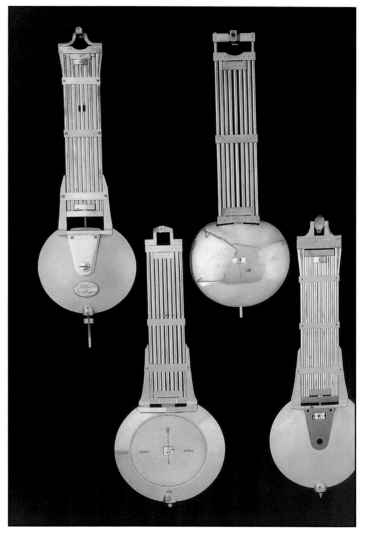

Figs. 5-15A, B. **A selection of nine rod gridiron pendulums** used on French table regulators dating from 1780-1860 with one incorporating an aneroid barometer in the center of the bob.

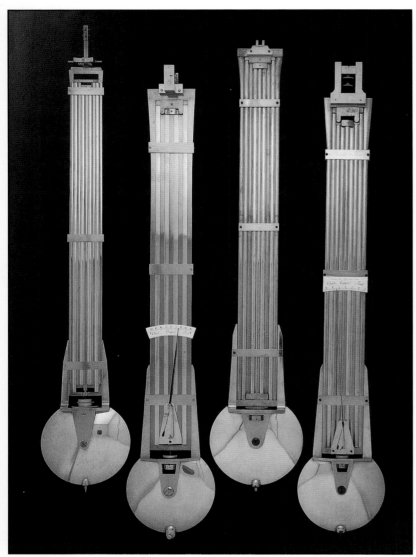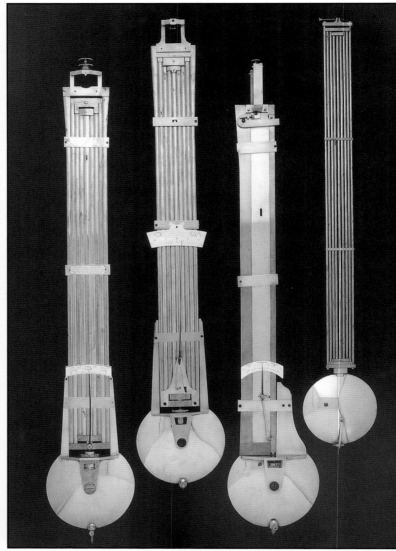

Figs. 5-16A, B. **A selection of French gridiron pendulums taken from longcase regulators.** Note the temperature indication on two of them and the complex form of some of the bars which have been taken out of one piece of steel. Berthoud's pendulum is second from the left, Fig. 5-16B. See also Fig. 5-18.

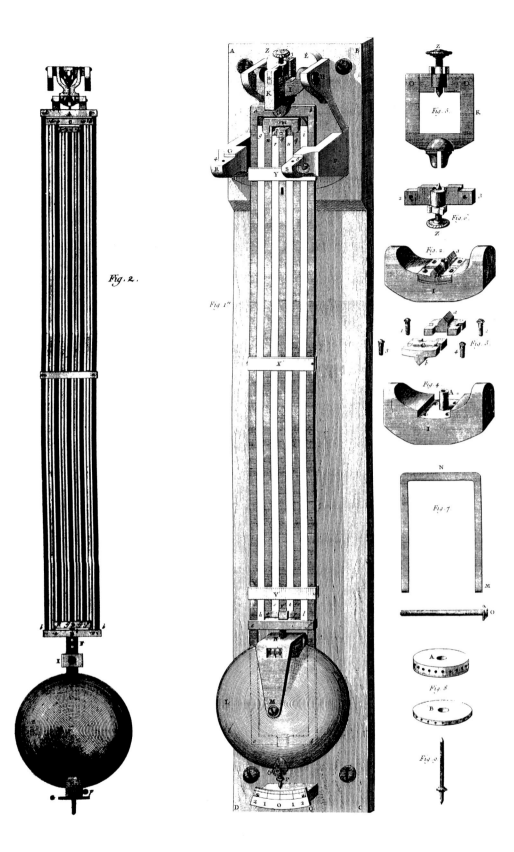

Figs. 5-17A, B. **Two designs by Berthoud for his nine rod gridiron pendulum.** 5-17A (left) originally appeared in Ferdinand Berthoud, Histoire de la Mesure du Temps par les Horloges, Paris 1802, planche IX and 5-17B (right) in Ferdinand Berthoud, Essai sur l'Horlogerie, Paris 1786, planche XXVIII. Both are reproduced, together with detailed descriptions, by Sabrier (vi).

Many of the French gridiron pendulums were masterpieces of design, their construction being made on a massive scale, and must have been extremely expensive to make. The rods, for instance, were frequently oval and often flared out and became wider at the top, which must have added appreciably to the work involved. Aesthetically they are very pleasing and most impressive, but it is unlikely that they performed any better than their English counter parts. Louis Berthoud spent much time on pendulum design and his thoughts and calculations, which go into considerable detail, are contained in his work books which still exist. These are reproduced in Jean-Claude Sabrier's excellent book on Berthoud.[11] Whilst his efforts were mainly confined to the nine rod gridiron pendulum (Fig. 5-17A, B) he also devised his own form of compensated pendulum which is seen in Fig. 5-18; however very few were made.

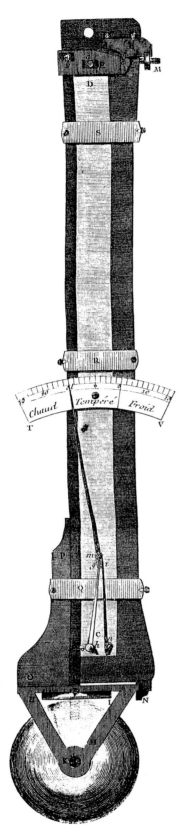

Fig. 5-18. **Berthoud's own design of compensated pendulum.** He abandoned this in favor of the nine rod gridiron after making only a small number. They are disconcerting to look at because, as they have an eccentrically mounted bob, they do not hang in a straight line. It has three rods, two of steel and one of brass. Because this results in inadequate compensation Berthoud designed an auxiliary lever, placed between the steel and brass bars. It appeared in Berthoud, 1763 Vol. 2 plate XXXIV, Vol.2, pp. 302 onwards. Fig. 5-16B shows a compensated pendulum, made to this design, probably by Berthoud, but fitted to a regulator by Delunesy.

Some idea of the thought he put into the design of his clocks and watches, and his pendulums in particular, is given by this brief extract from his entry of the 4th February 1796, taken from his third journal which appears in translated form in Sabrier.[12]

4th February 1796 - I have just calculated the dilation of my verge à chassis which has been made much too short and which I am obliged to lengthen by the support for the knife-edge, I should not hope to have an exact compensation, not at least unless the brass and steel used are of a quite different nature from that shown by the pyrometer:

The first grid of steel contains from the knife-edge	292 lignes
The second grid of 18 pouces 4 lignes	220 lignes
The third grid, to the centre of the bob	288 lignes
	800 lignes
which multiplied by the coefficient of the expansion of steel, 74, gives	59,200
The first grid of brass	241 lignes
The second grid of brass	220 lignes
	461 lignes
which multiplied by the coefficient of the expansion of brass, 121, gives	55,781

That is 59,200 - 55,781 = 3,149 which is lacking from the compensation and which makes about a nineteenth part, and to which must be added the error arising from counting in the first grid the part which carries the knife-edge as steel since it is brass. To make up something of this error, I had the length of steel added that was missing, that is 15 lignes. I note that this error of a nineteenth gives me about 4/360th of a ligne and should give an error of a little more than one second from cold to hot in 24 hours for an ordinary seconds clock and more than 2" for this.

The calculations continue for two more pages, gradually refining the design of the pendulum to produce the best thermal compensation. Similar notes refer to other regulators.

Berthoud took great care in measuring the relative expansion of various metals and the figures he obtained for these are given below; however, like others, he emphasized that the actual expansion of any component, such as a pendulum rod, can only be assessed accurately after it has been hammered up, machined, finished, polished and, where applicable, annealed.

Berthoud's Table of the Proportion of the Expansion of Metals, &c.

Steel softened	69	Silver wire drawn	119
Iron ditto	75	Brass	121
Steel tempered	77	Tin	160
Iron forged cold	78	Lead	193
Gold softened	82	Glass	62
Ditto wire drawn hard	94	Mercury	1235
Copper	107		

Ellicott's Compensated Pendulum.

The purpose of Ellicott's pendulum (Fig. 5-19), which he states he first designed c. 1732 but only presented to the Royal Society some twenty years later, was to try and overcome two possible problems with the gridiron pendulum; the difficulty of varying the degree of compensation being achieved and the tendency of the bars to move in a somewhat jerky fashion because of their liability to stick in the cross bars.

Fig. 5-19. **Ellicott's Compensated Pendulum.** The brass pendulum rod, which slides freely in front of a steel rod, presses down on the central shorter portions of two pivoted levers, the other ends of which support the bob. The points at which the bob is supported may be moved in or out by the graduated silvered screws to either side which are provided with pointers so that records may be kept.

He devised two methods of overcoming this, both of which are shown in his drawing in Fig. 5-20. With that on the left the bob rests on the longer ends of the two levers (Fig. 1 in the diagram), with the other ends being depressed and thus raising the bob when a brass bar expands more than the steel one to which it is linked.

The second method, based on the same principle and shown in Fig. 3 in the diagram, consists of a two foot long steel bar being fixed vertically to the backplate of the movement. Resting on the bottom of this and screwed to it through slots, so that it can move freely, is a brass bar. As the temperature rises the brass bar expands upwards more than the steel one will expand down and presses up on and lifts a horizontal bar pivoted at one end, and counter weighted at the other, to which the pendulum is attached. The exact point at which the brass bar presses on the horizontal one can be adjusted to vary the compensation. As the suspension spring passes through a close fitting fork this will vary its length and also its elasticity and thus the point from which the pendulum swings and therefore its effective length, which will compensate for any expansion taking place in the pendulum rod.

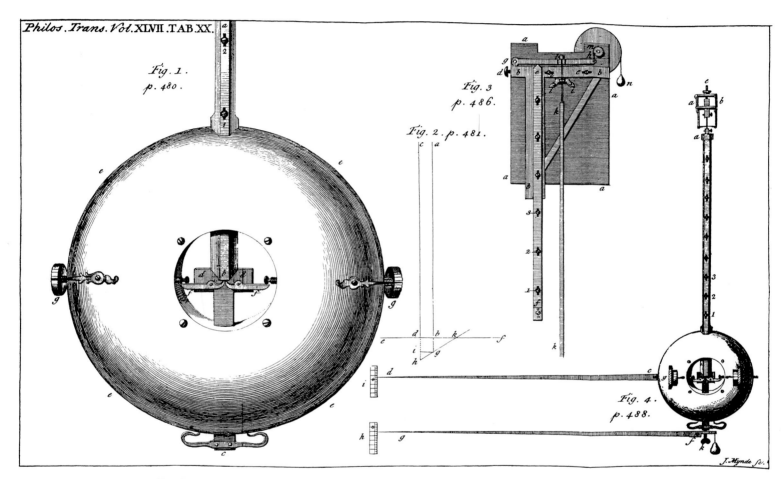

Fig. 5-20. **Ellicott's two methods of compensation**. Taken from his presentation to the Royal Society on 4th June 1752. Fig. 1 shows the first method in which a brass rod (a) moves freely up or down on a steel rod behind it, to which it is held by screws passing through slots 1, 2. The bottom of the rod (b) presses on the short inner ends of the levers (f f) the other ends of which raise or lower the bob. The amount the bob moves can be varied by moving the points of contact with the levers in or out by means of the screws g, g.

Fig. 3 shows his second method of compensation in which the bottom, f, of a brass bar rests on a steel bar to which it is freely attached, which extends down from the backplate. As the brass bar expands up it raises a horizontal steel bar, pivoted at g and counterpoised at the other end by the weight by means of a pulley. As the pendulum is lifted the effective length of the suspension is shortened and thus also that of the pendulum.

Fig. 5-21. **John Arnold & Son c. 1788.** The Ellicott style pendulum fitted to this clock incorporates Cumming's modifications, including the use of steel rods on either side of the brass one. There is a sliding weight on the rod for fine adjustment and also a small brass ball below the main bob for the same purpose.

Ellicott's original description of these forms of compensation, of which the former was by far the more popular, is contained in Chapter 12.

Alexander Cumming in 1766[13] suggested two improvements to Ellicott's pendulum. The first of these was the use of steel bars on either side of the brass one, rather than just the one, to prevent any tendency for it to flex, and ensure the free movement of the brass bar. The second was to modify the short ends of the levers and make them both act on the same point at the center of the brass bar immediately below the line of expansion (Figs. 12-20 D, E).

A pendulum based on Ellicott's original design, and incorporating Cumming's suggested modifications, may be seen in Fig. 5-21.

The French also used Ellicott's pendulum, the bob of a year clock by Mohren being seen in Fig. 5-22 and one by Lepaute in Fig. 5-23.

Fig. 5-22. **A massive pendulum with pivoted levers based on Ellicott's system of compensation.** The front of the bob, which is fully glazed to show the two bars to either side of the central one pressing down on the short ends of the pivoted levers, with the long ends supporting the bob. The regulation may be adjusted by using a key to move the support points for the bob in or out. The central gilt hand gives the amount of the temperature compensation. This pendulum comes from a year duration wall regulator by Möhren, Paris.

Fig. 5-23. **The pendulum from a regulator by Lepaute incorporating Ellicott's form of compensation** with pivoted steel levers supporting the bob, which are depressed by the two zinc rods. Temperature indication is also given.

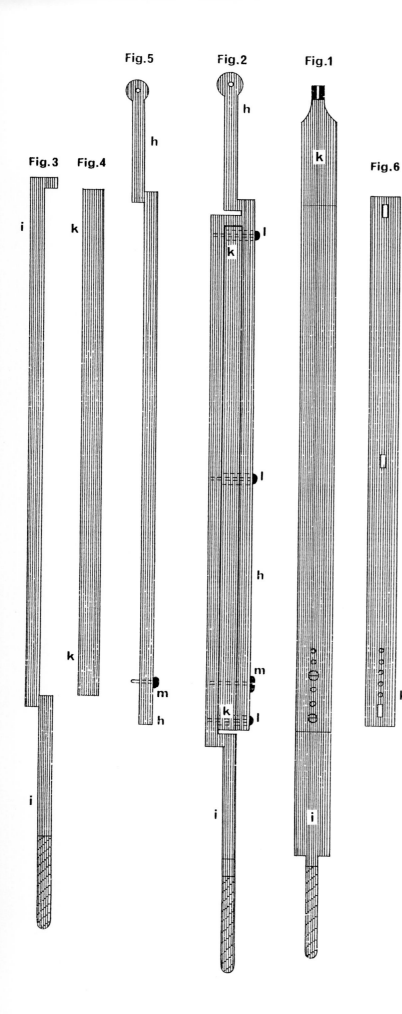

Zinc/Steel Compensated Pendulums Including those of Thomas Reid and Thomas Buckney

Zinc has been used in compensated pendulums instead of brass because of its much higher coefficient of expansion. Thus, whereas a combination of steel and brass is nearly ideal for a nine rod gridiron pendulum, if you wish to use only five rods then the expansion of brass will be inadequate and it has to be replaced by zinc, as in the case of Arnold's five rod pendulums.

The high coefficient of expansion of zinc also made other designs of pendulum possible such as that by Ward (Fig. 5-24)[14] which employs two flat bars of steel and one of zinc. Thomas Reid also includes further examples, but the best known of the early zinc/steel compensated pendulums is that which Reid himself designed (Fig. 5-25).

Fig. 5-24. **Ward's Compensated Pendulum**. The drawing of the pendulum rod and its components seen here have been redrawn from Reid[14]. There are two flat bars of steel ii and one of zinc kk which are held together loosely by three screws lll passing through slots so that they may move freely. The bar hh is connected to the bar kk by the screw m. The bar ii has a right angle shoulder on top which rests on the zinc bar kk. Figs. 3, 4, 5 are side views of the three bars and 6 shows the flat side of the zinc bar. Fig. 1 is a front view of the assembled rod.

The front steel bar hh, when it expands, carries down with it the central zinc bar kk which will expand up and take with it the other steel bar which carries the bob. Thus the central zinc bar compensates for the two steel bars.

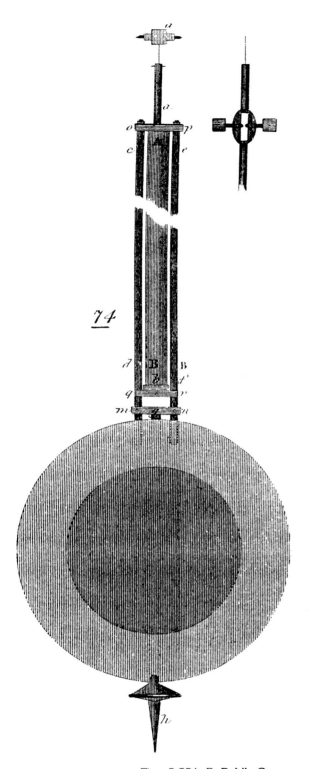
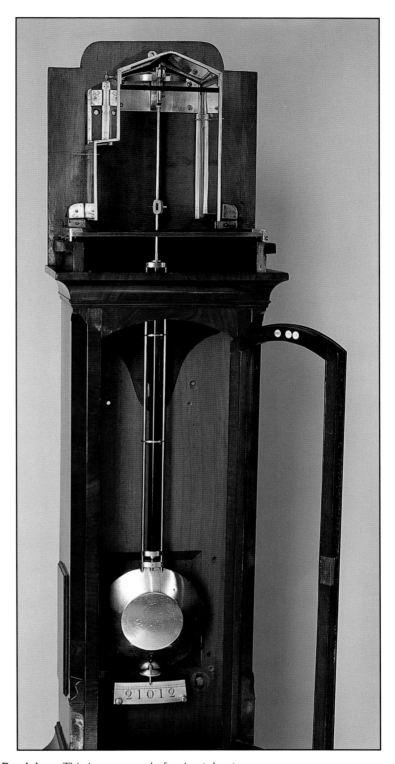

Figs. 5-25A, B. **Reid's Compensated Pendulum.** This is composed of a zinc tube, two outer steel rods and one central steel rod. As the central rod expands down with heat it will take with it the zinc tube that in turn will expand up and at the same time lift the top crossbar to which the outer steel rods are fixed. These expand down, taking the lower cross bar to which the bob is attached, with them. Thus the downward expansion of the inner and outer steel rods is compensating for the upwards expansion of the zinc tube.

Probably the most accurate was that fitted to Dent No. 1906 in 1870 and the one that was patented (No. 9358) by Thomas Buckney in August 1885 (Fig. 5-26). This used tubes of zinc and steel, both of which were perforated to permit the free passage of air within the tubes and thus enabled them to respond rapidly to any changes in temperature. An important part of the design, emphasized in the patent, is that the compensation could be varied by altering the point at which the upper ends of the zinc and steel tubes were connected.

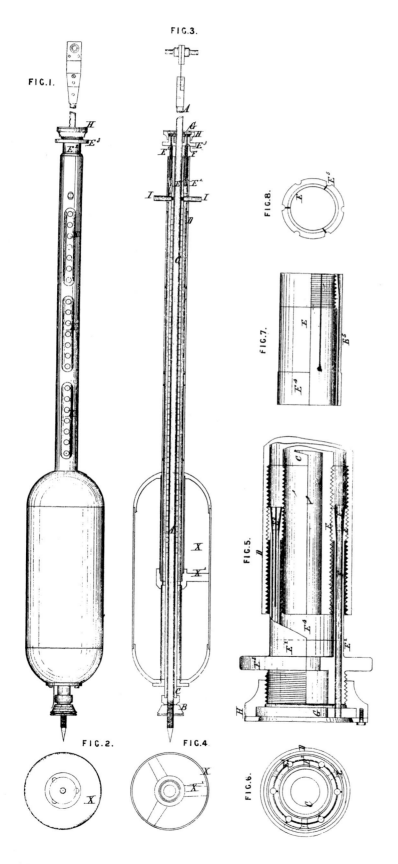

Left:
Figs. 5-26A, B. **Buckney's Zinc Steel Compensated Pendulum Patent No. 9358 dated 5th August 1885.** The important features of this pendulum are that the compensation is adjustable; the bob is supported from its mid-point and the tubes are perforated to speed any response to temperature change. Buckney's own description is given here:

"My invention relates to compensating pendulums wherein a steel pendulum rod is connected at its lower end to a tubular rod of zinc through which it passes, while the upper end of the zinc tube is connected to the upper end of an outer steel tube whose lower end is connected to the pendulum bob. My invention relates to means for adjusting the compensation of such pendulums by varying the point at which the upper end of the zinc tube is connected to the outer steel tube, whereby the ratio between the operative length of the zinc tube and the combined operative lengths of the steel rod and tube will be varied.

The construction which I employ for this purpose is shown on the accompanying drawings in which Fig. 1 shows an elevation of the pendulum, Fig. 2 a plan of the same, Fig. 3 a vertical section, Fig. 4 a cross section through the pendulum bob; Figs. 5 to 8 show enlarged details.

A is the steel pendulum rod, connected at its upper end to the suspending spring in the usual manner, and supporting on a screwed nut B on its lower end the zinc tube C, whose upper end is connected to the outer steel tube C, to the lower end of which is connected the pendulum bob X, by means of arms X^1 at the middle of its length. The mode of connecting the zinc tube C with the outer steel tube D will be readily understood on reference to Figs. 5, 6, 7 and 8.

The zinc tube C has a screw thread cut some distance down from its upper end upon its outer surface, and the steel tube D has a corresponding screw thread cut on its inner surface, and into these threads screws a cylindrical nut $E\ E^1$ having corresponding screw threads on its inner and outer surfaces, thereby connecting the two tubes together at a point which may be varied by screwing the nut up or down between them, this being effected by means of the cylindrical extension E^2 with milled flange E^3 projecting up beyond the ends of the tubes. Thus as the length of the zinc tube C is considerably less than that of the joint lengths of the steel rod A and tube D it will be seen that in altering the operative length of the zinc tube to the same extent as the operative length of the steel tube by shifting the position of the nut E E^1, the compensation will be varied. In order to ensure that the nut shall not shift its position when once it has been adjusted, I prefer to construct it as follows. The parts E and E^1 on which the inner and outer threads are formed, are made of two separate pieces, of which the inner part E shown detached at Figs. 7 and 8, has its extension E^4 connected at its upper end by screws or pins L to the extension E^2 of the part E^1. Both parts are split part of the way up at three or more points as shown, and they have cylindrical holes E^5 formed, half in the one part and half in the other, and into these holes fit rods F having conically enlarged ends F^1, the lower ends of the holes in the nut being correspondingly enlarged, so that on drawing the rods upwards their heads F^1 will force the part E inwards against the tube C and the part E^1 outwards against the tube D, thus jamming the nut tight between the tubes, while on lowering the rods again, the parts of the nut will spring back sufficiently to render it free to be adjusted.

The rods F are for this purpose connected by their upper ends to a ring G that is held loosely in a cap H screwed on to the upper end of the cylinder E^2, so that on screwing the cap upwards or downwards the nut is tightened or loosened as above described. The tube C is provided with two pins I I projecting through holes in the tube D by which it can be held so as to prevent it from turning when the nut E E^1 is being adjusted. In place of the rods F with coned heads any suitable equivalent means may be employed for effecting the required wedging action upon the nut. Holes are formed at K in both the tubes C and D in order to allow free access of the air to all three parts A, C and D, in order that they may be equally affected by change of temperature.

This pendulum was used by Dent on various observatory regulators and it appears to have given better results than the mercurial pendulum, probably because of its much more rapid response to changes in temperature. An example is to be seen in the year duration wall regulator (Fig. 20-36).

Another example of a zinc/steel pendulum was described by Hall[15] (Fig. 5-27) and a regulator incorporating zinc/steel compensation is seen in Fig. 5-28. The only disadvantage of incorporating zinc in compensated pendulums is that its long term stability is not as good as, for instance, brass or steel and it is more subject to corrosion.

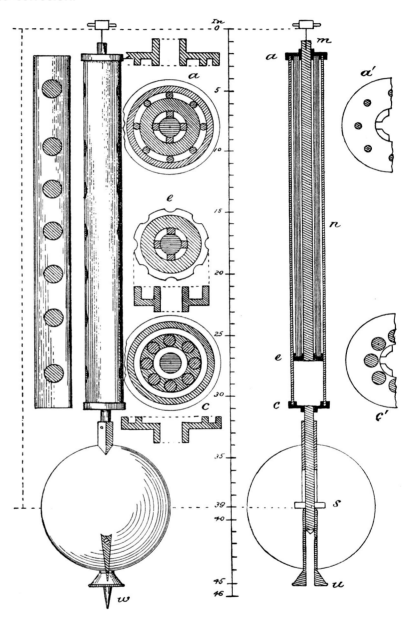

Fig. 5-27. **Hall's Zinc/Steel Compensated Pendulum.** Reproduced from the Horological Journal of November 1906, pages 37 - 39. The bob is supported at its center point by an extended regulating nut. Fitted to the bottom of the rod m is a disc e which slides freely within the outer tube n. This disc supports the inner tube that expands up and takes with it the cap a which slides freely on the central rod. Fixed to the cap by four screws is the outer tube n which carries the rest of the pendulum assembly including the bob. The cross sectional drawings give details of the inner and outer aspects of the two caps.

Figs. 5-28A, B **A high grade English Longcase Regulator, c. 1880** with zinc/steel compensated pendulum.

The Rhomboidal Pendulum

The origins of this pendulum go back to Hooke who suggested the use of a single rhomboid, the four sides of which were made of steel and could move freely at each corner with a central brass crosspiece which would, in effect, shorten the rhomboid by expanding outwards as the steel expanded down. This design had little practical application because the width of the pendulum would be as great as its' length: however a modification of the design was produced by Williamson of Ulverston, in Lancashire, c. 1760 which employed a series of five rhomboids one above the other, thus greatly reducing the width of the pendulum and these are occasionally encountered in clocks emanating from and around Lancashire.

Edward Troughton, at a somewhat later date, also produced a rhomboidal pendulum but with seven instead of five sections (Fig. 5-29). It will be seen that in this there are cross bars in each rhomboid whereas in the impressive French example of a rhomboidal pendulum (Fig. 5-30 A, B, C) just one zinc cross bar is used.

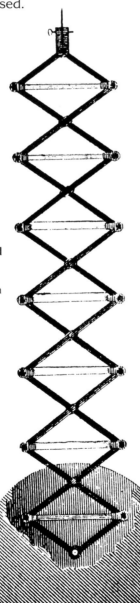

Figs. 5-29. **Troughton's rhomboidal pendulum** in which the frame is made of steel bars, all freely pivoted on each other, and cross bars, which as they expand, shorten the "concertina" and thus compensate for the downwards expansion of the steel.

Above and pages 94-95:
Figs. 5-30A, B, C. **French regulator of 80 days duration, c. 1820,** with finely executed articulated rhomboidal pendulum employing a substantial central turned bar of zinc. The movement is carried on two rectangular horizontal supports rigidly attached to a large brass plate. The front plate of the movement consists of a vertical square section screwed to the heads of the two horizontal support pillars. The 12lb. driving weight descends in a compartment behind the backboard.

Fig. 5-30B

Fig. 5-30C

Ritchie's Compensated Pendulum

David Ritchie laid his design for this (Fig. 5-31 and 32), together with an actual example, before the Adelphi Society in 1812 and as a result received a reward of twenty guineas.

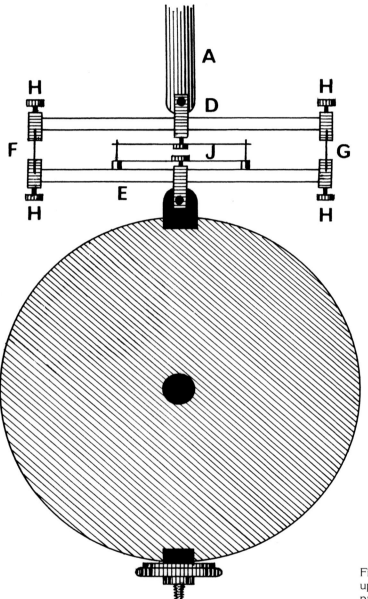

Fig. 5-31. **Ritchie's Pendulum.** D & E are two bimetallic bars with the upper surface of D and the lower surface of E of steel and the other parts of brass. Because the brass has a higher coefficient of expansion than steel, it will cause the bars, as the temperature rises, to curve inwards at the center and thus counteract the lengthening of rod A.

The bars are linked together by F and G into which the bars slide. By slackening off the locking screws H H these components can be slid in or out, thus varying the effective length of the bimetallic bars and the compensation being achieved.

To absorb some of the weight of the bob, two interlinked steel springs J A are screwed, one to the bottom of the rod and the other to the bob.

Fig. 5-32A, B. **Early 19th century longcase regulator by Thwaites & Reid, London with Ritchie's pendulum.** The movement has six heavy stepped pillars, wheelwork with six crossings, nine and twelve leaf pinions and end stops throughout the train.

The pendulum is based on Ritchie's but has one or two refinements such as screw regulation to move the components joining the bimetallic strips in or out to vary the degree of compensation.

The pendulum makes use of two bimetallic horizontal bars interposed between the rod and the bob. The upper part of the top one and the lower part of the bottom one are of steel with the inner strips being of brass. As the temperature rises the brass will expand more than the steel and the bars will curve inwards, thus shortening the distance between the rod and the bob and compensating for the expansion of the rod.

The only objection to this design is that, as the rod is impulsed by the crutch to one side, the inertia of the bob will cause stress and flexion of the steel strips connecting the two bars, which is obviously undesirable. Nicholson overcame this problem by moving the compensation to the top of the rod and eliminating the use of steel strips to link the horizontal bars.

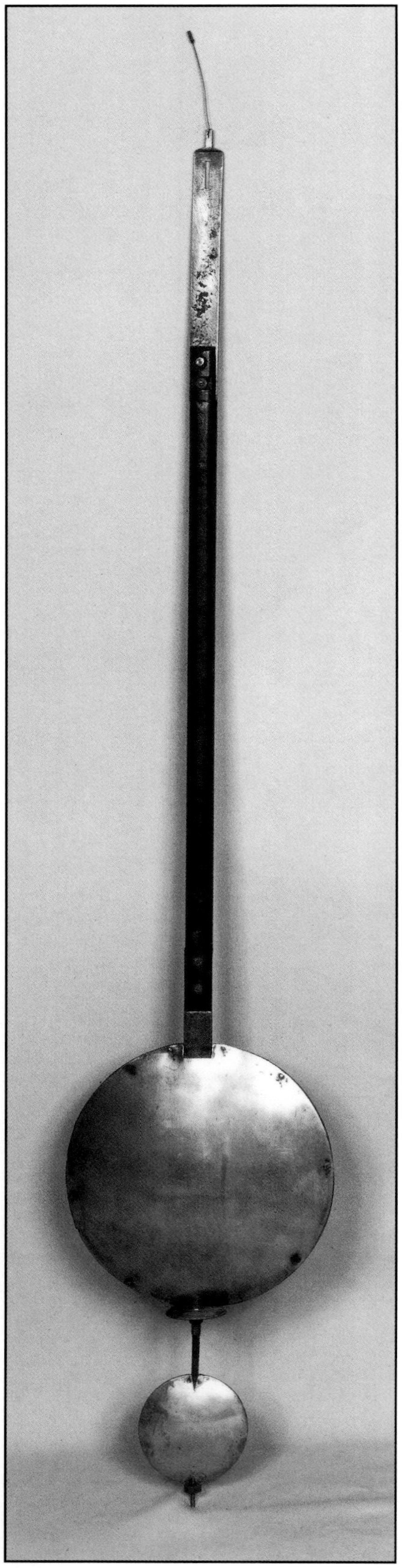

The Wood Rod Pendulum

The wood rod (Fig. 5-33) has been used extensively on regulators and also turret clocks over a very long period of time and the results achieved are often very good and yet many people, in some ways rightly, do not regard a wood rod pendulum as compensated.

Timber, like all materials, expands with rising temperature due to the increasing distance between its molecules as the magnitude of their oscillations increases. One of the difficulties is that the coefficient of expansion of different types of wood varies as of course do different metals, and that no two pieces of wood are exactly the same. It is thus important to select a particular type, pine being the most favored, with a maximum grain deviation of 1:60. This is important on three counts (a) wood expands and contracts far more (anything up to eighteen times) in cross than long grain, (Fig. 5-34), (b) it is less likely to distort, and (c) it is less likely to break.

		Coefficient of thermal expansion x 10^{-6}	
		Longitudinal	Transverse
Picea abies	Whitewood	5.41	34.1
Picea sitchsis	Sitka spruce	3.35	28.1
Pinus strobus	Yellow pine	4.00	72.7
Sequoia sp	Sequoia	4.43	29.5
Quercus robur	European oak	4.92	54.4

Fig. 5-34. **Coefficient of Linear Thermal Expansion of Various Woods Per Degree Centigrade.** Table provided by Professor J. Dinwoodie of the Center of Timber Technology and Construction.

However, the figures given in Fig. 5-34 tell only part of the story as the coefficients of expansion were all obtained at a constant humidity, whereas as temperature rises humidity normally decreases, which causes shrinkage of the wood which may well offset, or even exceed, the thermal expansion. Indeed in unsealed wood with a moisture content greater than around 3%, the shrinkage due to moisture loss with a rise in temperature will exceed the thermal expansion and thus the wood will shrink.[16]

The effect of the changes in humidity over a longer period of time also gives rise to considerable difficulties. We have all seen the dramatic effects and severe damage caused to fine antique furniture by lack of humidity due to central heating. Timber is hygroscopic and will absorb or release moisture into the atmosphere until it reaches equilibrium with it; thus if wood is held at a temperature of 25°C and 90% humidity until it has reached a stable state and then transferred to another chamber at the same temperature, but at only 60% humidity, its moisture content will drop from 21% to 12% with a corresponding decrease in length.[17]

A further factor to be taken into account is that the mass of the wood rod will change with humidity and thus the center of gravity of the pendulum as a whole will alter. With low humidity the rod will become lighter, thus lowering the center of gravity of the pendulum and slowing the clock and vice-versa.

The comments made so far assume that the wood rod is not sealed or protected in any way. If it is sealed, thus slowing both the ingress and loss of moisture, then the situation changes considerably and indeed in this case the pendulum rod may well expand to some degree with rises in temperature; however according to Heldman[18] although varnishing a pendulum rod would appear to reduce the moisture uptake by

Fig. 5-33. A wood rod pendulum with the main brass faced lead bob weighting some 14lbs. and a small 1/2 lb. subsidiary bob below it for fine regulation.

up to 60%, for reasons by no means apparent, it gives up that moisture far more slowly. Thus whereas an unvarnished rod .5" in diameter which had been subjected to 100% humidity for 48 hours gave up 2/3rds of the weight it had gained within 12 hours a varnished rod treated in the same manner still retained 25% of the weight it had gained two weeks later.

Coating of the exterior surface of wood does not affect the eventual quantity of moisture retained, it delays the time taken to reach equilibrium, varnishes retarding the rate of moisture exchange by 50-85% and paints by up to 90-95%.[20] Gilding was also used (Fig. 5-35).

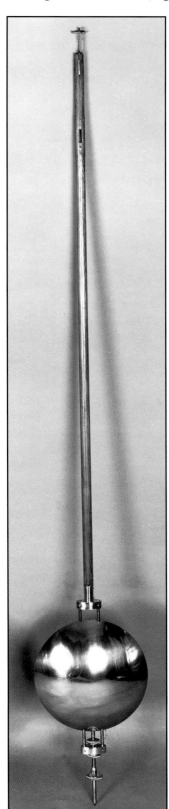

Fig. 5-35. A carefully conceived and executed wood rod pendulum in which the bob slides up and down on two steel rods and is supported at its mid-point by a central rod rising up from the regulating nut. The rod is gilded to reduce water absorption.

One last factor to be considered is the country in which the pendulum is going to be used, the moisture content of air dried wood varying, for instance, from 6.0-17.7% in Turkey and 16.8-20.3% in Iceland[19].

The foregoing will give some idea of the several imponderable problems attached to the design of the wood rod pendulum and the reason why Ferdinand Berthoud is recorded as having carried out experiments on it and discarded it.

Whereas it is undoubtedly true that for really accurate timekeeping pendulums such as the mercurial or gridiron, the design of which can be precisely calculated, are much to be preferred, it is likely that many regulators and also turret clocks with wood rod pendulums performed remarkably well, even if this was more by trial and error and above all the experience and knowledge of the clockmakers who had been making them over a long period of time.

At the time wood rods were used for pendulums only air dried timber, which would have been seasoned over a long period, was available. It is likely that this would have had a moisture content of around 22%, which may have dropped to around 18% indoors.[20]

Nickel/Iron (Invar) Compensated Pendulums

It was in 1896 that Dr. E. Guillaume, Director of the International Weights and Measures Office in Sèvres, outside Paris, discovered a nickel/iron alloy with a very low coefficient of expansion; just 1/12th that of steel; 1/18th that of brass and 1/23rd that of aluminum. The composition of the material initially tried was 35.7% nickel and 64.3% iron with small quantities of carbon and manganese. This, in developed form, was registered in France under the trade name of Invar.

The origin of this material is interesting in that it was not conceived for pendulums in the first instance.[21] For some time the Commission of Weights and Measures in France had been using an alloy of platinum and iridium to make their international standard meter lengths which obviously had to be as little affected by temperature as possible. These had proved very satisfactory but the cost was extremely high; Johnson, Mathey & Co charging 10,000 gold francs for each rod. It was for this reason that a less expensive alternative was sought. The first to be tried was a nickel/copper alloy, but problems were experienced with cracking and warping; however at that time a sample bar of ferro-nickel, 30% nickel with a little manganese and carbon, had been brought in to assess whether it might be used for weights. It was rejected as it was magnetic, but because it took a good polish and was free of kinks, Dr. Guillaume decided to study it. Much to his surprise he found it expanded less than any other known metal, and indeed only tungsten approaches it, but this was unknown at the time. After trying various alloys he came up with a ratio of 45% nickel, 0.40% manganese and 0.1% carbon, with iron 54.5%.

One of the first tests of Invar was in the balance of a chronometer being tested at Neuchâtel when it was found that the middle temperature error disappeared. Paul

Ditisheim was to repeat the experiment in two chronometers shown at the Paris Exhibition in 1900.

Although Dr. Guillaume continued to investigate the addition of various other metals none was found to be beneficial so far as pendulum rods were concerned. However, over a twenty year period, other alloys, such as Elinvar, were developed for hairsprings.

The potential of Invar was appreciated by Riefler at a very early period and by 1897 he had registered a patent, NDRP No 100870, incorporating this material; however it was realized quite early on that there were problems to be overcome if Invar was to realize its full potential.

The first of these was that the coefficient of expansion of the material varied by up to 100% between batches and so each had to be tested individually. The second problem was that it was by no means stable, tending to move in jerks with changes in temperature due to the release of molecular tension, which resulted initially in people getting poor results with it. This problem was appreciated by Dr. Guillaume within the first year. He continued to measure the stability of Invar rods over some twenty-seven years and found that during this period the growth was 50ppm, but that it gradually slowed down, some 35ppm taking place in the first six years.

Riefler overcame this problem, or at least mitigated it, by annealing the rod after it had been machined and finished. He used a special tempering oven in which twenty rods at a time were kept for twenty days. The temperature was raised to 120° C and this was reduced by 5° C each day with the rods being vibrated for a few seconds every four hours.

Because of the very low rate of expansion, compensation was far easier, tubes to a total length of 10cms, and of at least two different materials such as brass, aluminum or fused silica of variable length to suit the compensation required, being interposed within the bob between the point of support, at its center of gravity and the regulating nut (Fig. 5-36).

Not only the rod but the bob, and so far as possible all the fittings attached to the pendulum, were made of Invar. To ease the pendulum's passage through the air it, and in particular the bob, were finished up to a very high standard and then nickel or chrome plated to preserve the finish achieved, and indeed those on the J^2 pendulums were even for some time gold plated.

As with Riefler's mercury pendulum the coefficients of expansion, weight and density of all the pendulum's components were measured and from this their correct sizes, and particularly that of the compensating tubes, and the material to be used in them, was calculated.

An interesting account of the construction of Invar pendulums was written by Agar Baugh[22] in which many points were discussed such as the necessity of either using an Invar nut, which is difficult to make, not one of brass, on the Invar rod or to use a pin and then carry out fine regulation with weights. Other points were the optimum thickness of the rod and the design and fitting of the suspension spring.

His overall conclusion was the same as Riefler's; that the coefficients, etc. of all the components should be measured, an assessment made of any possible changes in the density of the air and then the problem passed over to a mathematician to determine the exact dimensions required.

Features of Riefler's design were:
1. A cylindrical bob when the clock was in a vacuum and a lenticular one in air.
2. The bob must be a loose fit on the rod to avoid any friction.
3. The adjusting nut should be of nickel-steel and have a square thread with a slope of 1 mm to eliminate any chance of it working down.
4. With his most accurate sidereal clocks a 14 mm rod was used with the adjusting screw having one hundred divisions. One turn altered the rate by forty seconds a day, i.e. 0.4 second per division.

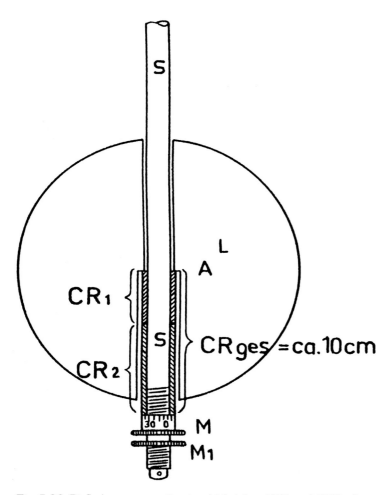

Fig. 5-36. Riefler's compensation in which tubes (CR1 and CR2) of at least two different materials, such as brass, aluminum or fused silica, with different coefficients of expansion were interposed between the regulating nut and support point of the bob at its center of gravity. The relative lengths of the tubes, of a total length of 10cms, were adjusted so as to give the correct amount of compensation.

5. For fine adjustment additional weights were used: (a) Three small ones of nickel silver, which would increase the rate by one second a day. (b) Three aluminum ones which make a difference of 0.5 seconds. (c) Five very small aluminium ones, each designed, when put on a tray half way up the pendulum rod, when it is in air, to affect the rate by 0.1 second. (d) If the pendulum is in a vacuum then the tray should be only 27cms below the center of oscillation.

For ultimate performance barometric compensation was fitted (Fig. 5-38). The result of this attention to detail was that the maximum final error per day per °C was 0.005 seconds.

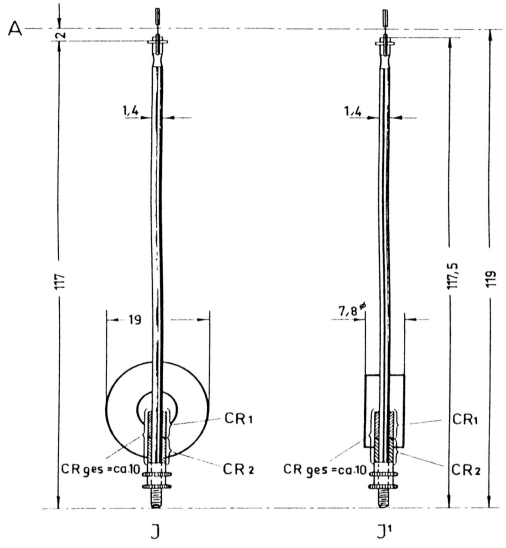

Fig. 5-37. The dimensions of Riefler's pendulums, left used in free air and right in a tank regulator unit.

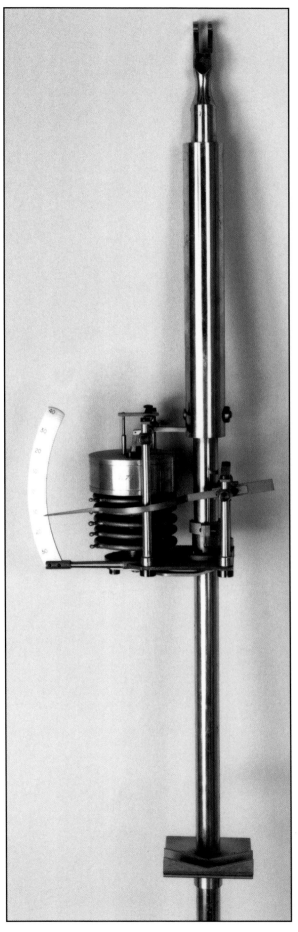

Fig. 5-38. Riefler's barometric compensation unit with stratified temperature compensation above it.

101

With the invar pendulums, one hundred were made from a single block or ingot, but despite this their coefficients of expansion, varied quite widely after they had been manufactured, and thus each one had to be tested at the International Bureau of Weights and Measures at Sèvres or the Imperial Standardization Commission in Berlin.

Standard invar was used by Riefler from 1897-1938; super-invar from 1938-1955 and fix-invar for first class pendulums from 1955-1958. Super-invar had a quoted coefficient of 0.8×10^{-6} per °C; from -50°C to +100°C and with fix-invar the figure was reduced to 0.5×10^{-6} per °C. With this material, to improve stability, the impurities were reduced and a small amount of chrome added.

A problem which Riefler had perceived at a comparatively early stage was that of stratification, i.e. the change of temperature, for instance within a room, with height. This becomes obvious if, for instance, you climb a ladder in a heated room in the winter. It is far warmer nearer the ceiling. When related to accurate timekeepers this means that the temperature, and thus the rate of expansion, will increase from the bottom to the top of the pendulum; the problem being exacerbated when electric heaters are used. By 1809 Riefler was offering a solution to this difficulty (Fig. 5-39).

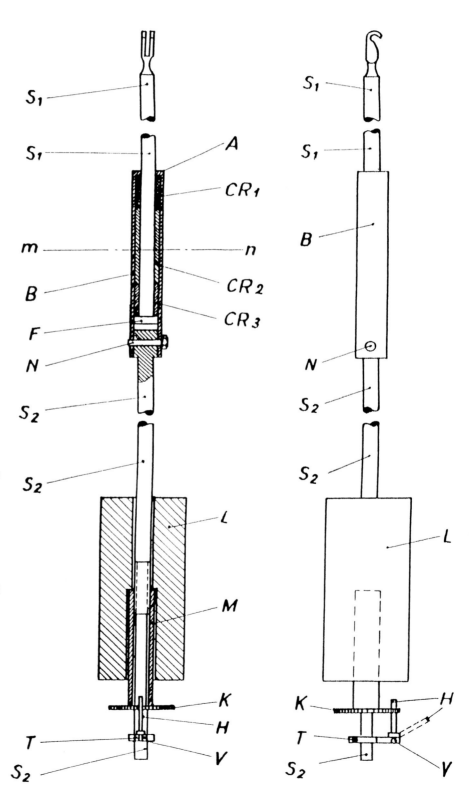

Fig. 5-39. Riefler's compensation for stratified temperature, i.e. compensation for the variations in temperature at different levels in the room and thus the pendulum rod also.

The upper part of the pendulum rod S, has a flange F at the lower end of which the compensation tubes CR1, 2 and 3 rest. The lower part of the rod which carries the bob L is bolted (by N) to the tube B, the top end of which A, rests on the upper compensation tube and will thus be raised or lowered by it. The compensation tubes are so designed that the lengths above and below their mid point m - n are the same.

The bob L rests at its center of gravity on the top of the regulating nut M which is made of the same material as the bob. To prevent any chance of rotation flat threads are used. K is the regulating disc used to rotate it and H the hinged locking lever.

The compensation mechanism was moved from below the bob to the upper half of the pendulum rod. An upper and a lower portion were used with the exact position required being determined after measuring the coefficients of expansion of the rod, the bob and the suspension spring. This will vary depending on whether the pendulum is in air or a vacuum. Approximate figures are 47.5 cms below the pivot point for a vacuum and 21 cms in free air. An average figure for the J^{sch} pendulum when fitted with aneroid compensation is 16 cms

Some 230 pendulums with compensation for stratification were made and, in all, some 3,839 Nickel-Steel pendulums, many being sold to other regulator manufacturers.

Riefler's pendulums are also discussed in Chapter 33.

Because of the extremely low coefficient of expansion of quartz and its excellent stability, Riefler did investigate the possibility of using this material and even made and tested one pendulum incorporating it. This gave a similar result to super-invar; however due to the difficulty of manufacture, its fragility and electrostatic attraction to dust, it was never produced. Ceramic quartz, which has virtually zero expansion, was also investigated but abandoned for the same reasons as ordinary quartz glass.

The tale of Invar is not yet finished. Right up until the present time it is still being used in pendulum clocks and its various derivatives discussed. Robert Matthys [23] produced a good review of the subject in 1995. In this he comments on the instability of the material and the more recent attempts to overcome this, such as the reduction in the impurities. He also discusses one of the later forms of invar, that has been made easier to machine, a problem in the past, by the addition of 0.2% of selenium. Unfortunately this material increases in length with time, roughly twice as fast as regular invar.

The Quartz Pendulum Rod

Although glass was occasionally used for the pendulum rod in Viennese wall clocks and regulators by Waugh of Dublin and Bannister of Colchester (Fig. 17-9), this was of relatively little advantage as it has a fairly high coefficient of expansion and is, of course, fragile. Quartz or fused silica (silicon dioxide) on the other hand, although equally fragile, has three advantages (a) its coefficient of expansion is extremely low; Brinkmann[24] giving figures of 0.000001 per °C for invar and 0.0000004 for quartz (b) it expands very evenly without the jerks associated with invar and (c) it is a very stable material and thus its coefficient of expansion does not change with age as does that of invar.

Probably the biggest disadvantage of using quartz for the pendulum rod is the difficulty of making any attachment to it for the suspension or the bob, it being virtually impossible to cut a thread for a nut in it; however, shellac[25] will provide a good bond and nowadays we can use cyanoacrylate glue or epoxy resin.

It is believed that the first quartz pendulum (Fig. 5-40A, B) was devised by Frederic Ecaubert of New York (US Patent No. 965507, issued on 26th July 1910). In his patent he emphasized the fragility of the material and the difficulty of supporting the bob and fixing the pendulum hook. He overcame this problem by not fixing them directly to the quartz tube.

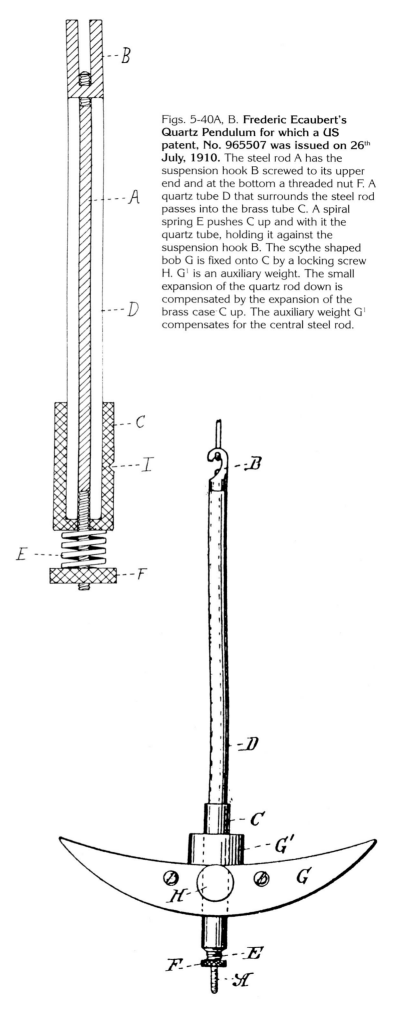

Figs. 5-40A, B. **Frederic Ecaubert's Quartz Pendulum for which a US patent, No. 965507 was issued on 26th July, 1910.** The steel rod A has the suspension hook B screwed to its upper end and at the bottom a threaded nut F. A quartz tube D that surrounds the steel rod passes into the brass tube C. A spiral spring E pushes C up and with it the quartz tube, holding it against the suspension hook B. The scythe shaped bob G is fixed onto C by a locking screw H. G^1 is an auxiliary weight. The small expansion of the quartz rod down is compensated by the expansion of the brass case C up. The auxiliary weight G^1 compensates for the central steel rod.

Besides the quartz pendulum shown in Fig. 5-40, Ecaubert designed a second with two parallel rods which is illustrated and described by Gerhard Streitberger[26] in his excellent review of quartz pendulums.

Adolf Herz of Vienna patented the use of quartz (Austrian Patent No. 62446) for pendulums on 3rd April 1912 but gave only a brief description of their design. Some three months later Satori also patented a quartz pendulum (Fig. 5-41A, B, C) but went into considerable detail regarding its construction. (Austrian Patent specification No. 62448. Submitted 10th June 1913 issued 10th Dec 1913). In this he made use of screws to fix both the suspension and the bob to the rod. A particularly interesting feature was the use of a removable device (Fig. 5-41C) for raising and lowering the bob.

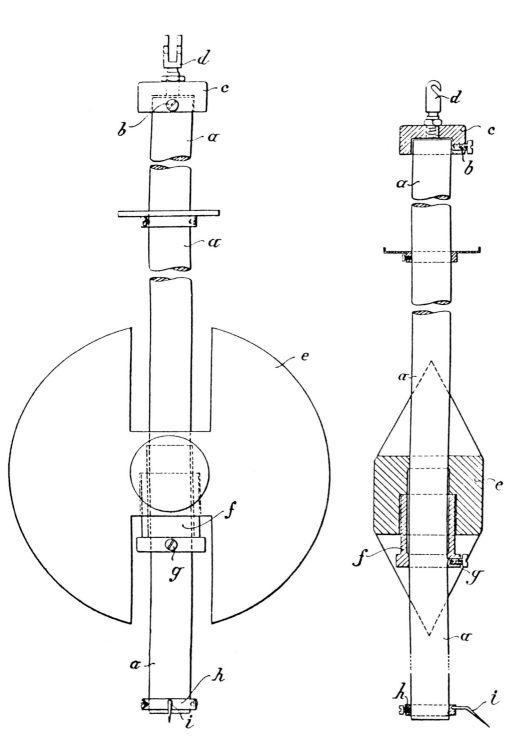

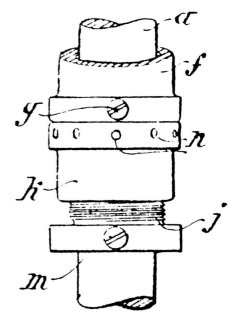

Figs. 5-41A, B, C. **Satori's Quartz Pendulum, patent No. 6248, granted 10th Dec. 1913.**
The top of the quartz rod "a" is fixed to a metal cap "c", to which the suspension hook "d" is fixed, by means of a screw "b." The bob rests on the upper edge of f which is fixed to the quartz rod a by screws g. The ring h at the bottom carries the pointer for the beat scale. A tray is provided for timing weights.

In 6C the removable apparatus for regulating the pendulum is shown. The two components j and k screw into each other with the lower part j being fixed to the pendulum rod by locking screws. To raise or lower the bob after removing the ring h, the appliance is pushed up the pendulum rod until the upper component k is close to the bottom of f. j is now fixed to the pendulum rod by the screws m and the screws g are then slackened off. By rotating k with a tommy bar inserted in the holes n, n, fine adjustment can be carried out. When the regulator is to time and screws g are tightened again and those holding j in place slackened so that it can be removed.

Further fine adjustment can be carried out by adding or removing small weights from the tray on the rod.

In 1912 Reubold also produced, but did not patent, a 3/4 second pendulum with quartz rod[27] (Fig. 5-42). Riefler also produced and tested a pendulum with quartz rod and found that it performed as well as his nickel-iron pendulums, but he did not pursue it, probably mainly because of its fragility, bearing in mind that his regulators were frequently shipped abroad.

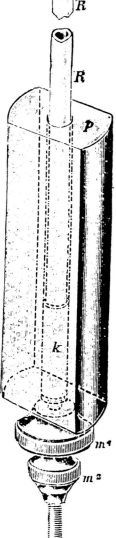

Fig. 5-42. **Reubold's 3/4 seconds quartz pendulum, produced in 1912.** At the top is a tube F to which the quartz tube R and the hook h are fixed. To the bottom of the quartz rod K is fastened. This has a threaded section below onto which the regulating nuts m^1 and m^2 are screwed. Resting on the regulating nuts is the flattened iron bob, the height of which is carefully calculated to compensate for the quartz rod. The pendulum is impulsed by a counter weighted pin touching one side of the rod close to the tube clamp g.

The quest for the perfect pendulum still continues to this day, with quartz in the forefront, although the modern ultra low expansion glasses may now produce the better result, however they are much more expensive. Helpful articles on this have been written by E.C. Martt[28] and R.J. Matthys.[29]

One modern material worth mentioning briefly here is carbon fiber which has a very low thermal coefficient and apparently good long term stability; however it is somewhat susceptible to changes in humidity.[30]

The ways devised for compensating for the effects of temperature on the length of a pendulum are almost endless and we are concluding this part of the chapter with one of the most ingenious, that of Nicholson (Fig. 5-43).

Above and following page:
Fig. 5-43A, B, C. **Nicholson, London (1735 - 1815).** This table regulator that is glazed to the top and sides, achieves its thermal compensation by raising or lowering the pendulum by means of changes in the curvature of the bi-metallic strips from which it is suspended. The effective length of the pendulum is dictated by the steel plates impinging on either side of the suspension spring.

Nicholson's own form of gravity escapement is fitted which is described in Chapter Six and shown in Fig. 6-10. It was devised by him in 1784 and this clock is dated 1797.

Fig. 5-43B

Fig. 5-43C

The Regulation of Precision Clocks

When the timekeeping of clocks was relatively poor, the coarse regulation achieved by means of a nut raising or lowering the bob was adequate but as their performance improved so finer methods of adjustment had to be employed. Initially large diameter regulating nuts with a fine thread were used with divisions marked and numbered on them. These were sometimes graduated for precise periods to time; thus Riefler provided a regulating nut with forty teeth on its periphery on some of his pendulums which could be locked in place by a hinged lever (Fig. 5-44). A full turn altered the rate by forty seconds a day and a single tooth by just one second. A refinement employed with wood rod pendulums in particular until well into the nineteenth century was the use of a small bob below the main one (Fig. 5-33) which could be brought into play after the timekeeping had been got fairly close with the large bob.

Fig. 5-44. Riefler's regulating nut with forty divisions which could be locked in place by a hinged lever. Each division represented a change in rate of one second a day. Note the adjustable beat plate below the bob.

George Graham, on his mercurial pendulum, used an auxiliary weight which could be slid up or down the pendulum rod. An alternative to this is the placing of a metal tray on the pendulum rod to which small weights can be added. This has the advantage that the pendulum does not have to be stopped. Martin Burgess[31] has calculated that with the tray 1/3rd of the way down the rod from the effective point of suspension, approximately 13", if you add 1/10,000 of the weight of the pendulum to the tray it will change the rate by one second a day.

A refinement on this theme was that used by Riefler (Fig. 5-45) in which one weight, to which a cord is attached, rests permanently on a tray on the pendulum and the other is suspended just above it. These were normally adjusted so that taking off or putting on a weight, which can be done without stopping the pendulum, produced a change of rate of + or - 0.1 sec per hour.

Fig. 5-45. **Riefler's method of bringing a clock to time.** One weight, to which a cord is attached, rests on the tray whilst the other is held just above it. Putting on or taking off a weight varies the rate by 0.1 sec. per hour.

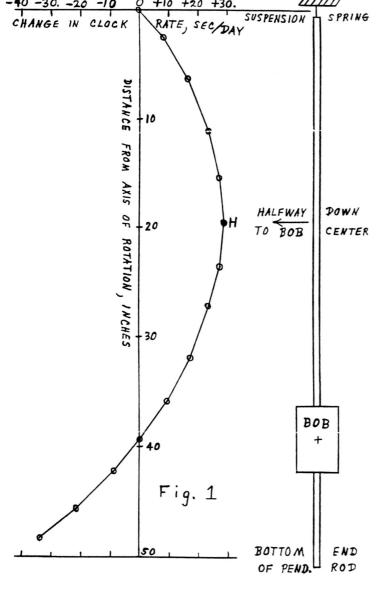

Robert Matthys did interesting research into (a) the effect of adding a weight at various different points along the rod of a pendulum, including below it and (b) the effect of moving a sliding weight up or down the pendulum rod. The results are seen in Figs. 5-46 and 5-47.

Fig. 5-46. This graph, produced by Robert Matthys, shows the effect of adding a small weight at various positions along the pendulum rod. The maximum effect occurs at the half way point H, with the clock going faster if the weight is added above the bob and slower if it is below, although the exact effect will be determined to some extent by the relative masses of the rod and bob.

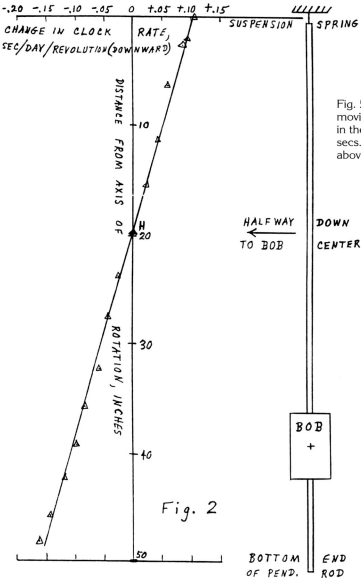

Fig. 5-47. This graph, also by Robert Matthys, shows the effect of moving the weight along the rod, as was done by various clockmakers in the past. A 24 gm. weight produced a total change of rate of 0.097 secs. per day. Moving the weight down the rod speeds the clock if it is above the half way point (H) but slows it when it is moved below this.

With astronomical regulators in particular, it was highly desirable that the clock could be regulated without stopping it. Christopher Wood[32] describes an interesting modification to a clock by Reid and Auld that achieved this purpose. The effective length of the pendulum could be varied by a known constant amount that would advance or retard the time of the clock by one second in forty minutes. By employing this device on the regulator, which was used for time service, it could be brought back to exact time each day.

With the advent of the movement which was fully enclosed within an airtight case, it became a comparatively simple matter to regulate the clock by varying the barometric pressure (Fig. 5-48).

Fig. 5-48. This table, contained in Riefler's instructions on how to set up a tank clock, shows the relationship between changes in pressure and changes in rate, so that the clock may be regulated or brought to time by this means.

The changes in pressure in column I correspond to the daily changes in rate in column II.

The readings of the barometer inside the air-tight case can be compared with one another only when they are made at the same temperature. When the temperature rises, the air inside the air-tight case expands and the pressure rises. Depending on the evacuation of air within the case, the pressure rises between 0 and 2.78mm. per degree Celsius.

I m/m	II seconds	I m/m	II seconds	I m/m	II seconds	I m/m	II seconds	I m/m	II seconds
1	0.018	21	0.378	41	0.738	61	1.098	81	1.458
2	0.036	22	0.396	42	0.756	62	1.116	82	1.476
3	0.054	23	0.414	43	0.774	63	1.134	83	1.494
4	0.072	24	0.432	44	0.792	64	1.152	84	1.512
5	0.090	25	0.450	45	0.810	65	1.170	85	1.530
6	0.108	26	0.468	46	0.828	66	1.188	86	1.548
7	0.126	27	0.486	47	0.846	67	1.206	87	1.566
8	0.144	28	0.504	48	0.864	68	1.224	88	1.584
9	0.162	29	0.522	49	0.882	69	1.242	89	1.602
10	0.180	30	0.540	50	0.900	70	1.260	90	1.620
11	0.198	31	0.558	51	0.918	71	1.278	91	1.638
12	0.216	32	0.576	52	0.936	72	1.296	92	1.656
13	0.234	33	0.594	53	0.954	73	1.314	93	1.674
14	0.252	34	0.612	54	0.972	74	1.332	94	1.692
15	0.270	35	0.630	55	0.990	75	1.350	95	1.710
16	0.288	36	0.648	56	1.008	76	1.368	96	1.728
17	0.306	37	0.666	57	1.026	77	1.386	97	1.746
18	0.324	38	0.684	58	1.044	78	1.404	98	1.764
19	0.342	39	0.702	59	1.062	79	1.422	99	1.782
20	0.360	40	0.720	60	1.080	80	1.440	100	1.800

References

[1] *Rees's Clocks Watches and Chronometers 1819-20.* pp. 237. Taken from The Cyclopaedia; Universal Dictionary of Arts, Sciences and Literature by Abraham Rees. Reprinted by David Charles 1970.

[2] Riefler, D. *Riefler Präzisions-Pendeluhren von 1890 - 1965.* pp. 36 - 38. Callwey 1991.

[3] *Riefler Mercurial Pendulum.* German Patent No. 60,059. 20.03.1891.

[4] Finn, J.L. & Riefler, S. *Compensating Pendulums and how to make them.* Hazlitt &Walker. Chicago. 1905.

[5] Stewart, A.D., *The accurate compensation of Graham's mercurial pendulum.* Antiquarian Horology Spring 1996. pp. 416 - 421.

[6] King, A. L. *John Harrison, Clockmaker at Barrow, Near Barton Upon Humber; Lincolnshire. The Wooden Clocks, 1713-1730*, included in The Quest for Longitude. Andrewes, W. J. H. Harvard University, 1996.

[7] Laycock, W. S. *The Lost Science of John "Longitude" Harrison.* Brant Wright Associates Ltd 1976.

[8] Burgess, M. *The Scandalous Neglect of Harrison's Regulator Science'.* Included in The Quest for Longitude. Andrewes, W. J. H. Harvard University, 1996.

[9] Andrewes, W.J H. *John Harrison, A Study of his Early Work.* Horological Dialogues, Vol. I, pp. 11-38, 1979.

[10] Reid, T. *Treatise on Clockmaking.* pp. 362 - 367. Blackie & Son, Glasgow, Edinburgh and London (4th Edition) 1849.

[11] Sabrier, J.C. *La Longitude en Mer à L'heure de Louis Berthoud et Henri Motel.* Mainly pp. 160 - 175.

[12] *ibid*, pp. 160 - 161.

[13] Cumming, A. *The elements of Clock and Watch-work, Adapted to Practice.* In two essays, 1766, article 334 - 343 and plates 9, 10 & 11, and p. 106 from article 359.

[14] Reid, T. *Treatise on Clock & Watch Making.* Blackie & Son, Glasgow, Edinburgh and London. Fourth edition 1849. Plate XVI.77 and pp. 376-378.

[15] Hall, J.J. *A New Arrangement for Zinc and Steel Compensated Pendulums.* Horological Journal November 1906, pp. 37 - 39.

[16] Dinwoodie, B. Centre for Timber Technology & Construction. Personal communication.

[17] Tsournis, G. *Science & Technology of Wood.* p. 155. Van Nostrand Reinhold, New York.

[18] Heldman, A. *Effect of Humidity on Wood Pendulum Rods.* Horological Science Newsletter. April 2000. P.24-27.

[19] *Op. cit.* Tsournis. p. 138. Table 9-1.

[20] Hedgecock, John. British Woodworking Federation. Personal Communication.

[21] Guillaume, C.E. Invar. *A lecture given at the AGM of the British Horological Institute.* December 1930, pp 70 - 72 & January 1931, pp. 83 - 86.

[22] Agar Baugh, J H. *Invar Pendulums.* Horological Journal, pp. 46 - 48 November 1926 & pp 69-72 December 1926.

[23] Matthys, B. *Instability of Invar.* Horological Science Newsletter, December 1995 pp 3-6.

[24] Brinkmann, H. *Einfükrung in die Uhrenlehre.* Düsseldorf 1974. p. 72.

[25] Irwing, R. *A fused silica pendulum rod.* Horological Science Newsletter. April 1996.

[26] Streitberger, G. *Zur Erfindung des Quartzpendels.* Alte Uhren 5.10.88. pp. 85-90.

[27] Streitberger, G. *Op. Cit.* p.89.

[28] Martt, E.C. *Notes on Fused Silica & Ule.* Horological Science Newsletter. 1996. Issue 5. p.14.

[29] Matthys, R.J. *Some Practical Properties of Quartz.* Horological Science Newsletter. 1997 Issue 2. p.9.

[30] Edwards, E. *The effect of moisture in a carbon fibre pendulum rod.* Horological Science Newsletter. March 1996.

[31] Burgess, M. *Temperature Compensation.* Horological Journal. April 1984. pp.13-14.

[32] Wood, C. *The Astronomical Clocks at the Observatory, Carlton Hill, Edinburgh.* Antiquarian Horology. Dec. 1972 pp. 55-62.

Chapter 6
Escapements
BY JOHN MARTIN

In attempting to describe the development of the escapement from the mid seventeenth century, when serious efforts were first made to achieve accurate timekeeping, it seemed logical to look at events in chronological order. From that time until the early part of the twentieth century many hundreds of designs for escapements for pendulum clocks have been published. We have attempted, in the ensuing pages, to describe those used in pendulum clocks which have obviously contributed to the development of accurate timekeeping, with occasional passing reference to interesting variations or, perhaps concepts, that, had they been taken up by practical clockmakers, might have proved valuable. Of necessity much has had to be left out of this chapter, and we can only apologize to those readers who can find no mention of a particular design that they feel should have been included.

We can define an escapement as a mechanical device that will allow power stored in a clock train to escape at a controlled rate, governed by the action of a balance or a pendulum to which it imparts impulse. The first known examples appeared somewhere in the thirteenth century in the form of a "verge and foliot". This device consisted of an escape or crown wheel with an odd number of teeth driving the flags or pallets of a vertically mounted staff. This in turn was suspended on a cord and carried at the top a crossbar, known as a "foliot", with notches in its top edge and, in turn, carrying two weights whose positions could be adjusted to regulate the timekeeping

For house clocks the crude device of the foliot was fairly quickly replaced by a balance wheel, and this then would be a form of escapement that would remain in common use for hundreds of years. In fact, the verge balance was still in use for cheap watches late into the nineteenth century.

The application by Huygens, in the mid 1650s, of the pendulum directly to a clock train was to create major

Fig. 6-1. **Miniature table portico clock, signed Franciscus Swartz in Bruessel c. 1630 fitted with Burgi's cross beat escapement.** The principle is to use a single escape wheel with two horizontal pallet arbors. These two arbors are geared together and carry their own balance arms. The action causes the arms to cross and re-cross, hence cross-beat

changes in the concept of timekeeping. The verge escapement was retained, but suffered one small complication—the necessity for the crown wheel to be mounted on a vertical staff required the train to be turned through a right angle using a contrate second wheel. Initially the verge was fitted with a crutch and the short pendulum was suspended on a thread that hung between cycloidal cheeks, as Huygens appreciated the advantage of their use to maintain isochronism.

The London clockmakers were quick to adopt the use of the pendulum and, whilst those who were making to high standards maintained the use of the separately suspended pendulum for some time, most attached it rigidly to the verge itself. However, it is interesting to note that the French never favored this latter course, and almost invariably continued to suspend their pendulums separately and drive them via a crutch.

The combination of the verge and short pendulum continued in use for bracket clocks and certain Continental wall and table clocks until the end of the eighteenth century. However, the timekeeping qualities of such an arrangement left much to be desired, and very early in the application of the pendulum a most important development occurred with the arrival of the anchor escapement.

Before considering this important step though, let us go back briefly in time. Writing in 1673, the astronomer Hevelius referred to a highly accurate escapement which appeared to consist of two foliots arranged crosswise, but it was not until this century that more detail came to light, when Professor Hans Von Bertele discovered a series of clocks with, what he termed, a cross beat escapement. Through his subsequent researches it seems clear that this escapement was invented by Jost Burgi, (1552–1632), and the design does seem to have been a major advance on anything that had gone before. (Fig. 6-1).

Although this design could have been exploited it was soon overtaken by the arrival of the pendulum and, M.J.L. Kesteven[1] suggests that Burgi, in this design, came near to creating a pendulum controlled escapement, had he extended one arm of the balance sufficiently. To return to the anchor escapement. The wide arc of swing taken up by a pendulum impulsed by a verge escapement, (approximately 40 deg.), necessitated the use of a short pendulum. However, after the publication by Huygens of his cycloidal theory it was realized that, to be isochronous, a pendulum should swing in a cycloidal, rather than a circular arc. It, therefore, follows that with a simply suspended pendulum, the smaller the arc the less the cycloidal error, and this leads to longer and heavier pendulums. To follow this course through required a re-think of escapement design.

In 1669 Dr. Robert Hooke, an eminent London physicist, set up an experiment with a 13 foot, 2 seconds' pendulum, which was given impulse near its lowest point by a pin in the periphery of a watch balance. This demonstrated the minute amount of power required to keep a heavy pendulum in motion provided that the amplitude was small; in this case only about 0.5 deg. In addition such a pendulum is isochronous because, over such a small arc, the circular and cycloidal curves are, to all intents and purposes, coincidental.

On the strength of this experiment it had long been thought that Hooke was the inventor of the anchor escapement. However, some now think that its invention should be credited to William Clement who, from about 1675, made longcase clocks using a 60 inch, 1-1/4 second pendulum on a spring suspension. (Fig. 6-2A) There is evidence, though, that Joseph Knibb had already made clocks with a rudimentary form of anchor escapement whilst still working in Oxford, i.e. prior to 1670.[2]

Fig. 6-2A. **Movement of a Clock by William Clement.** This clock has a 1-1/4 seconds' pendulum and is fitted with quarter striking. It is shown without the strike-work and the left half of the frontplate to show the anchor escapement. Whilst this is a later replacement it is interesting to note what is said about this clock in Ronald Lee's book, *The First Twelve Years of the English Pendulum Clock*, where it appears as item 24. The following is a precis of the relevant parts of the description:

"*...the movement reveals the fact that it was marked out for a verge, but the plates not actually drilled. Further examination confirms that drilling took place to accommodate a cross-beat escapement. One must assume that this was unsatisfactory in the narrow case in which this movement is housed, as its oscillation needs additional width. It is, therefore, fitted with an anchor escapement and because some additional width was required a 61 inch pendulum was used, so enabling Clement to obtain the extra width by placing the pendulum bob within the base.*"

It is then said "*Acceptance of this theory places this clock as No.1 anchor escapement and long pendulum clock.*"

Since the publication of this book in 1969 much research has been carried out into the origins of the anchor escapement, and it is probably unreasonable to accept this conclusion but the clock remains a very fine example of Clement's early work.

Another example of a very early anchor escapement is seen in the timepiece by Joshua Winnock, illustrated in Figures 6-2B, C. (Also see *Early English Clocks*).[3]

Fig. 6-2B. **Escapement of a very early timepiece signed Joshua Winnock Londini Fecit**. This very early clock is in its original narrow ebonized case. The 63 inch pendulum beats 1-1/4 seconds and maintains a very narrow arc of swing. The escape wheel and pallets are original and clearly show the individual thought that went into the infancy of the anchor escapement.

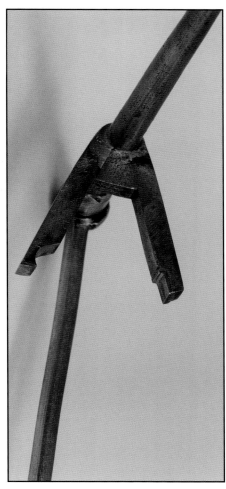

Fig. 6-2C. **Pallets of the Joshua Winnock clock.** These are original and have merely had their acting faces built up over the intervening years to eliminate wear.

We shall probably never know who thought of it first, but the anchor escapement was such a revolutionary concept that it completely changed the face of mechanical timekeeping, and with it the science of the Precision Pendulum Clock was born. Fig. 6-3 illustrates the conventional anchor escapement arranged to drive a seconds' pendulum.

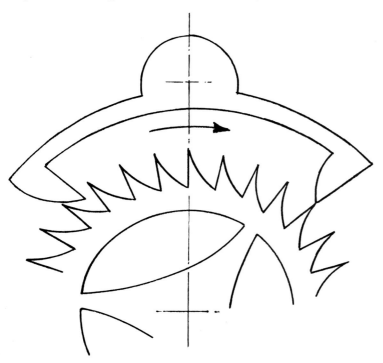

Fig. 6-3. **Conventional Anchor Escapement for a Clock with a seconds' pendulum.** Note that the geometry of this design is different from those in Figs 6-2A, B and C, because of the shorter pendulum.

It seems generally accepted that the first serious attempt to produce a precision pendulum clock was by Tompion when, in 1674, he was commissioned to make the two clocks for the new Royal Observatory at Greenwich.

The first Astronomer Royal, John Flamsteed (1646-1719), was a close friend of Richard Towneley of Burnley, who contributed much to late seventeenth century scientific investigation and was described as "an ingenious philosopher and a useful contributor to the Royal Society." It was he, Towneley, who it is thought designed the escapements for these clocks.

Tompion looked back to Hooke's experiment five years earlier and used a 13 foot, 2 seconds' pendulum, which was given impulse below the bob by a crutch from the escapement. The clocks were of year duration with a pendulum having a total arc of around one degree, and would have been, effectively, isochronous.

The escapements of these clocks did not, however, have recoil pallets. It must have been quickly realized that the recoil represents lost power and, although this matters little for normal house clocks, when looking for high precision, the reduction of friction throughout the train becomes an important factor.

The invention of the anchor escapement, therefore, created the platform on which subsequent development of most mechanical escapements was based for the next one hundred and eighty years. The principal example of this development being known as a dead beat escapement, of which the two Tompion clocks used early examples.

The dead beat is sometimes said to have been invented by George Graham around 1715. Graham had been Tompion's business partner until the latter's death in 1713, and, until his own death in 1751 he specialized in making fine watches, and, although his output of clocks was not as great as that of his erstwhile partner, his work made a much greater contribution to precision timekeeping, and maintained, (some would say surpassed), the standard of quality that Tompion had set.

The claim, however, for his invention of the dead beat must now be dismissed. In addition to earlier examples there exists a fine sidereal and meantime month duration regulator signed Tho. Tompion, London, fitted with what appears to be an original dead beat escapement and which can be reasonably safely dated to 1708-1709.[4]

Together, therefore, with the recent researches into the Tompion clocks at Greenwich, there is no doubt that the dead beat was around much earlier than had been thought. Indeed Howse, in his article in Antiquarian Horology 1970-1971[5], points out that Graham never claimed credit for inventing the dead beat and suggests that, in future, he be credited, not with the first example, but with the first successful one. Even this claim however, is challenged by the considerable amount of research that has been undertaken since Howse published his article.

The dead beat, (or Graham escapement as it is often known), became by far the most commonly used and successful mechanical escapement for high class regulators right up to the present day. It is reasonably easy to make and it is efficient, requiring a light driving weight and maintaining a consistent performance over long periods between maintenance (Fig.6-4). However, in common with most mechanical escapements, it suffers from the necessity to maintain lubrication of the pallets and many makers have tried to overcome this problem.

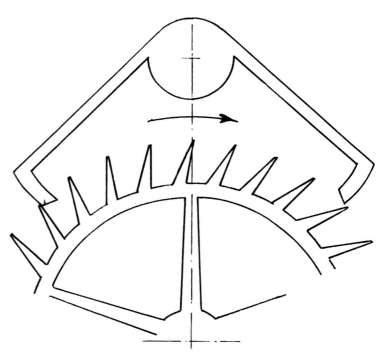

Fig. 6-4. **Graham Dead Beat Escapement**

The first, undoubtedly, were the Harrison brothers. John and James Harrison were carpenters living in Barrow-on-Humber in Lincolnshire, and they were commissioned, around 1722, to make a clock for Brocklesbury Park nearby. Built substantially of wood, the movement is going to this day and has the first example of their grasshopper escapement. (Fig. 6-5A)

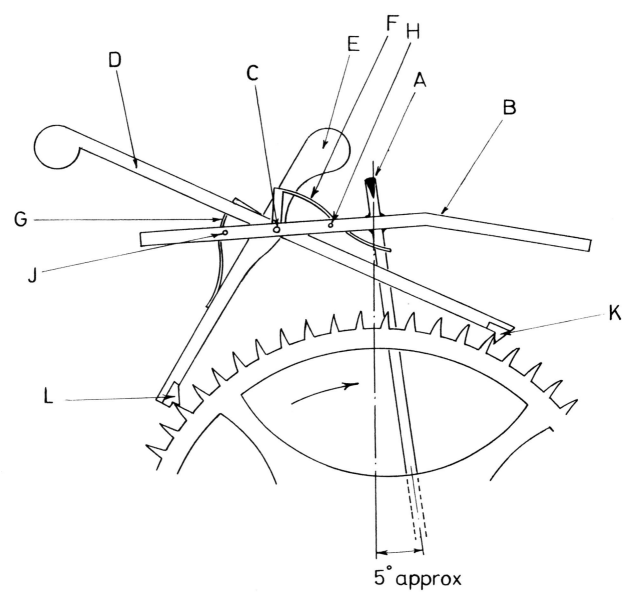

Fig. 6-5A. **Harrison's "Grasshopper" Escapement.** The pendulum is suspended on knife edges at **A** and, attached rigidly to it, is a brass frame **B** in which, freely pivoted on a pin **C**, are the two pallet arms **D** and **E**, and two pieces that Harrison referred to as 'Composers' shown as **F** and **G** in the drawing. The composers are made of brass and they rest on pins **H** and **J** respectively. The pallet arms are very light, being made of wood and are tail-heavy, the pallets are shown as separate but originally they were in one piece with the oak pallet arms.

In the illustration both pallets are shown engaged by escape wheel teeth. The shape of the teeth is important as it is necessary that they lock into the corner of the pallet, there being no sliding motion in this escapement. The pendulum is moving to the right and has nearly reached its maximum semi-arc of 6deg. In the last degree of movement the pendulum applies a small downward pressure on the escape wheel via the pallet **L**. This causes a slight recoil of the escape wheel, in turn releasing the locking at **K** and the tail-heavy pallet flies upwards to be restrained by its composer **F**.

The pendulum then commences its swing to the left whilst the pallet **L** remains engaged and the pallet arm **D**, resting against the face of its composer, moves down to engage the next tooth.

By the time the brothers moved to London in the mid 1730s they had made several longcase clocks using the concept of wooden movements employing lignum vitae bushings and lantern pinion rollers requiring no lubrication.

Reid, in his Treatise[6], quotes a Professor Robison as having a conversation with John Harrison and reporting as follows:

Having been sent for to look at a turret clock, which had stopt, he went to it, though it was at a considerable distance from his home, and found that the pallets were very much in want of oil, which he then applied to them. On his returning, and ruminating by the way on the indifferent sort of treatment which he thought he had met with, after having come so far, he set himself to work to contrive such an escapement as should not give to others that trouble to which he had been put, in consequence of this turret clock.

Whilst Graham strived, with his design, to avoid recoil at the cost of friction and the necessity for lubrication, the Harrisons' concept incorporated some recoil and minimized friction. One might consider that here was an excellent escapement for precision pendulum clocks, but the design dictates a requirement for a pendulum arc of around 12deg. necessitating, as recognized by the brothers, the use of cycloidal cheeks and a very accurate train to minimize circular error.

Although a very elegant design, the grasshopper suffered seriously unless great care was taken when adjusting the clock; the light, fragile pallet arms being easily broken. Very few examples remain to display its use by makers other than the Harrisons. Notably, though, Benjamin Vulliamy made some fine regulators using a different form of Harrison's grasshopper escapement. (Fig. 6-5B).

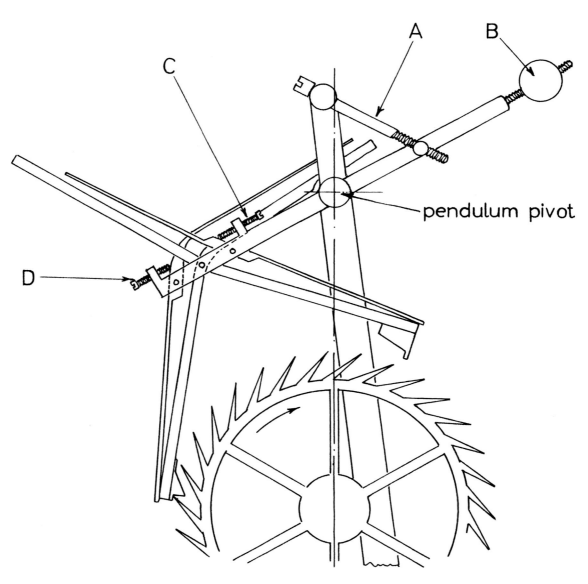

Fig. 6-5B. **Benjamin Vulliamy's version of Harrison's "Grasshopper."** This is a considerable variation of this escapement. The pallet arms are of ivory. The control frame is not rigidly fixed to the pendulum, but pivots about the latter's suspension point and is held rigid by an adjusting screw **A**. This, together with the adjustable weight **B**, enables the escapement to be put into beat. Unlike Harrison's original design the composers are pivoted separately from the pallets, and their position can be adjusted by the stop screws **C** and **D**. The sharp and angled escape wheel teeth would also improve the lock into the pallet angle which is an important feature of the "Grasshopper."

As stated earlier, the importance to precision horology of Graham's escapement cannot be over-emphasized, but it still suffered from the problem that only the Harrisons had solved—the need for lubrication. As the viscosity of the oil changes so the power transmitted to the pendulum will vary.

A number of clockmakers had observed that it is an essential requirement of a successful Graham escapement that the lightest possible driving weight be used. Should the weight be increased the pendulum amplitude tends to decrease. This is a function of additional friction on the dead faces of the pallets and can be exacerbated by the lubrication problem. Having said all this, however, the advantage of this simple and trouble-free design far outweighs the disadvantages.

As the middle of the eighteenth century approached much thought was being given to accurate timekeeping in France. In the 1740s a Parisian clockmaker, named Amant, produced a variation of the Graham escapement utilizing pins projecting from the escape wheel face, rather than teeth around the periphery. Originally round pins were used and projected only from one side of the wheel (Fig. 6-6A).

This design was quickly improved with the upper half of the pins being removed in the manner of the pallet pins of a French pin pallet escapement. This contributed to a reduction in the drop of the pin onto the pallet, (see Fig. 6-6B), but it was the eminent clockmaker, Jean Andre Lepaute, who contributed most to its improvement. He placed each pallet on opposite sides of the wheel and arranged the pins to project alternately from each side of the wheel on different pitch circles (Fig.6-6C).

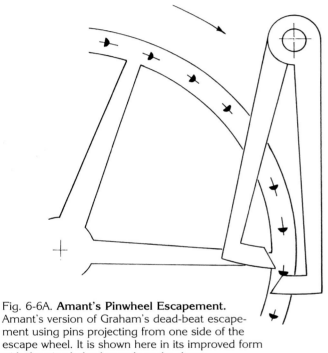

Fig. 6-6A. **Amant's Pinwheel Escapement.** Amant's version of Graham's dead-beat escapement using pins projecting from one side of the escape wheel. It is shown here in its improved form with the pins halved to reduce the drops.

Fig. 6-6C. **Lepaute's version of the Pinwheel Escapement.** Two sets of pins are used projecting alternately from each side of the wheel, having their drop-off points arranged on the center line.

Fig. 6-6B. Amant's Escapement fitted to the backplate of, unusually, an English regulator by Henry Walsh of Newbury. Note the counterpoise required when the pallets are not set vertically.

Lepaute's design is an excellent escapement. However, it requires twice the number of pins of Amant's, and they are, of necessity, smaller in diameter. In addition they are quite often left round for obvious practical reasons, with the consequent loss of efficiency due to increased drop.

A further variation of the pinwheel escapement is to be seen in the Coup Perdu, which was used in half second pendulum spring driven table regulators to enable them to display seconds. (Fig. 6-6D).

Fig. 6-6D. **Coup Perdu (lost beat) pinwheel Escapement.** This is a clever adaptation of the pinwheel used by some French makers of table regulators, with half second pendulums, to overcome the problem of indicating seconds.

The right hand pallet is conventional but the left hand arm carries a small detent which is pivoted. The escapement is shown locked on this detent but, as the pendulum moves to the left, the escape-wheel pin drops onto the impulse face of the right hand pallet. As soon as it is free, the pivoted pallet rotates anticlockwise due to the small weight projecting above it, the amount of movement being controlled by two stop screws.

The pendulum then begins its movement to the right and the escape wheel rotates anti-clockwise giving impulse. The locking detent on the left slides under the next pin and is pushed downwards by it until it again rests on its stop screw in the position shown.

The pinwheel design continued in use for fine French (and sometimes English) regulators, being treated as an alternative to the Graham, and another celebrated clockmaker, Robert Robin, in the late eighteenth century developed a version for watches. He also went some way to overcoming the lubrication problem with its use in clocks as, when the radial faces of the pallets lie in the horizontal plane, he cut shallow grooves in their dead faces to function as oil baths.

Antoine Thiout (known as Thiout L'Aine) was born in Paris in 1692 and became a celebrated clockmaker. In 1741 he published his *Traite d'horologerie* in which he describes many different concepts of escapement (including Amant's pinwheel hot from the press), some of which are re-workings of earlier designs. A principal example of such is one invented by Chevalier de Bethune. Thiout says that he, Thiout, first applied it in 1727 and that it was adopted by the majority of clockmakers who knew about it. (Fig. 6-7A, B).

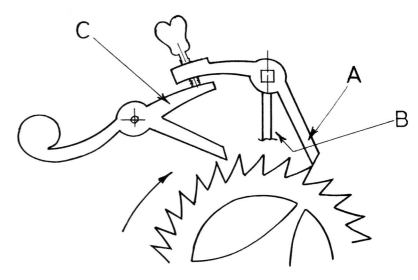

Fig. 6-7A. **Thiout's Separate Pallet Escapement.** The exit pallet **A** is rigidly fitted to the pendulum crutch **B** whilst the entry pallet **C** is separately pivoted and has a counterpoised pallet arm to make it tail-heavy. An adjusting screw is provided to set the pallets in depth.

Fig. 6-7B. **An example of the use of Thiout's design in a small French regulator movement.** Despite Thiout's claim that it was adopted by the majority of clockmakers who knew about it, very few examples seem to have survived to the present day.

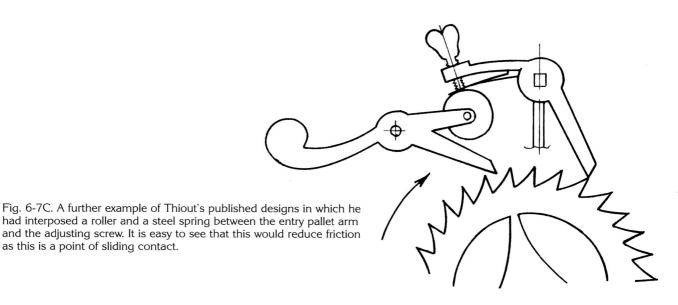

Fig. 6-7C. A further example of Thiout's published designs in which he had interposed a roller and a steel spring between the entry pallet arm and the adjusting screw. It is easy to see that this would reduce friction as this is a point of sliding contact.

In his *Traite* Thiout follows on with a number of variations of his own on the same theme, one of which is shown here in Fig. 6-7C which, he says, gave improved performance, and certainly one can see that the use of the roller would reduce friction. Thiout's work is well described by Chamberlain.[7]

As we enter the second half of the eighteenth century we find that the principal developments in precision timekeeping are taking place in France and Britain. In France, the nine rod gridiron compensated pendulum is in general use and, in Britain also it was employed for most observatory regulators, possibly in part because of its ease of transport. However, by the early nineteenth century the gridiron had been largely replaced by the mercury compensated pendulum, a pattern which was to be repeated in France as the century progressed.

The leading clockmakers were building great precision into their pendulums, but they were all dogged by the necessity to connect them to a dead beat escapement directly through the medium of a crutch, thus transmitting to them the inevitable errors due to variations in power through the train. Their minds, therefore, turned more and more to the possibility of isolating the escapement from the pendulum.

Constant Force or Remontoire Escapements

What are we trying to do in an escapement? We are using a mechanism to transmit power to a pendulum in order to maintain its movement. However, the ability of a pendulum to beat time is a natural characteristic, and it was realized very early in the development of the pendulum clock that the greater the reduction in power variation through the train, the less the interference with the motion of the pendulum.

Much thought was given to methods of isolating the train from the escapement and/or the escapement from the pendulum whilst maintaining adequate impulse to keep the latter in undisturbed motion. We can, therefore, broadly classify all subsequent development of the mechanical escapement under the heading of Constant Force. Such escapements are also known as Remontoire Escapements.

The concept of the remontoire had been used by many French clockmakers, since the mid eighteenth century, as an attempt to maintain constant force on the escape wheel of a spring driven clock. The basis of the idea is that the train terminates in a fly and does not run constantly. At predetermined intervals it is allowed to run and rewind a tiny weight or tension a small spring that powers the escape wheel.[8] Such a device is normally known as a Train Remontoire, and high quality French regulators have used this method since the mid eighteenth century to provide power to a dead beat escapement, either pinwheel or Graham. However the idea of building the remontoire into the escapement itself was now beginning to exercise the minds of clockmakers.

Perhaps the first successful attempt to produce such an escapement was by Alexander Cumming, an Edinburgh born clockmaker who worked in London, and, in 1766, published his *Elements of Clock and Watchwork*.[9] In this he describes a gravity escapement of some ingenuity utilising separately pivoted pallets, each operated by a small weight. (Fig. 6-8)

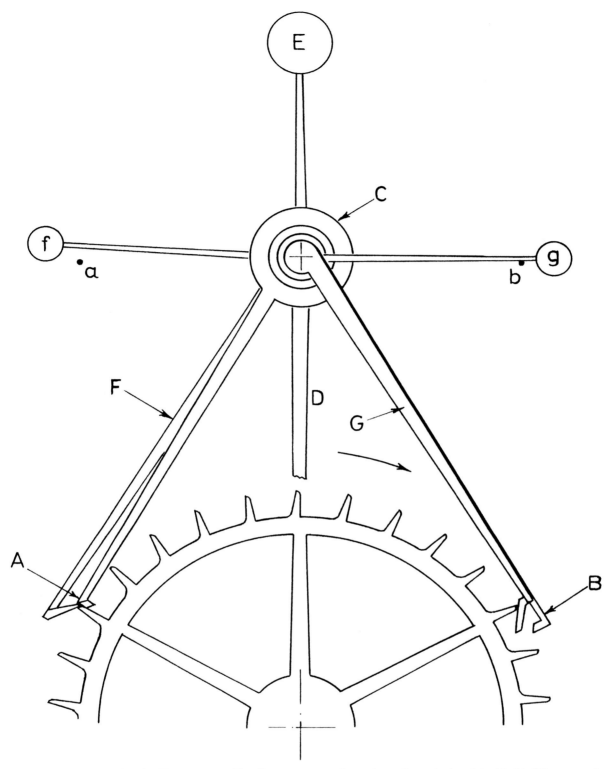

Fig. 6-8. **Cumming's Gravity Escapement.** The illustration is re-drawn from Cumming's original in his 'Elements of Clock & Watchwork'

The two long arms carrying the locking detents **A** and **B** are rigidly attached to a cylinder **C**, which also carries the crutch **D**, this whole assembly being counterpoised by the weight **E**. Also carried on this assembly is a cross bar (not shown) which has two pins **a** and **b** at its extremities, and the cylinder **C** is pivoted front and rear.

Pivoted separately and independent of each other within the cylinder are two pallet arms **F** and **G** and each has a long arm with a weight **f** and **g** attached to it. (It should be noted that, in the drawing, the pallet arm **G** virtually obscures the arm of the detent **B**, whereas the arm of detent **A** is in front of pallet arm **F**).

As seen, the escapement is locked on the detent **A** and the pallet arm **F** is fully raised. If we move **D** to the left the pin **a** will rise towards the arm of weight **f**, and at the same time the escapement will unlock and the wheel will turn clockwise lifting the pallet arm **G**, and relocking on detent **B**. As the pendulum then moves to the right it will receive impulse from the weight **f** acting on the pin **a**. The action will then be repeated on the other side.

Cumming put this design into practice in a magnificent clockwork barograph commissioned by George III in 1765, which is still in the Royal Collection (Fig. 12-19), and followed it with another built for himself (Fig. 12-20).

Some two years after Cumming published his work, Thomas Mudge also produced a design for a gravity escapement along similar lines. Insinuations were made, at the time, that Mudge had "borrowed" his design from Cumming, but Mudge had, in fact, used the principle in 1763, some two years before Cumming had built his barograph, and applied it to a balance for improving marine timekeepers, using springs instead of weights to operate each pallet separately. Mudge's design is much simpler than Cumming's in that the pallets and detents are combined: the detents being in the form of small nibs at the extremities of the pallets. (Fig. 6-9).

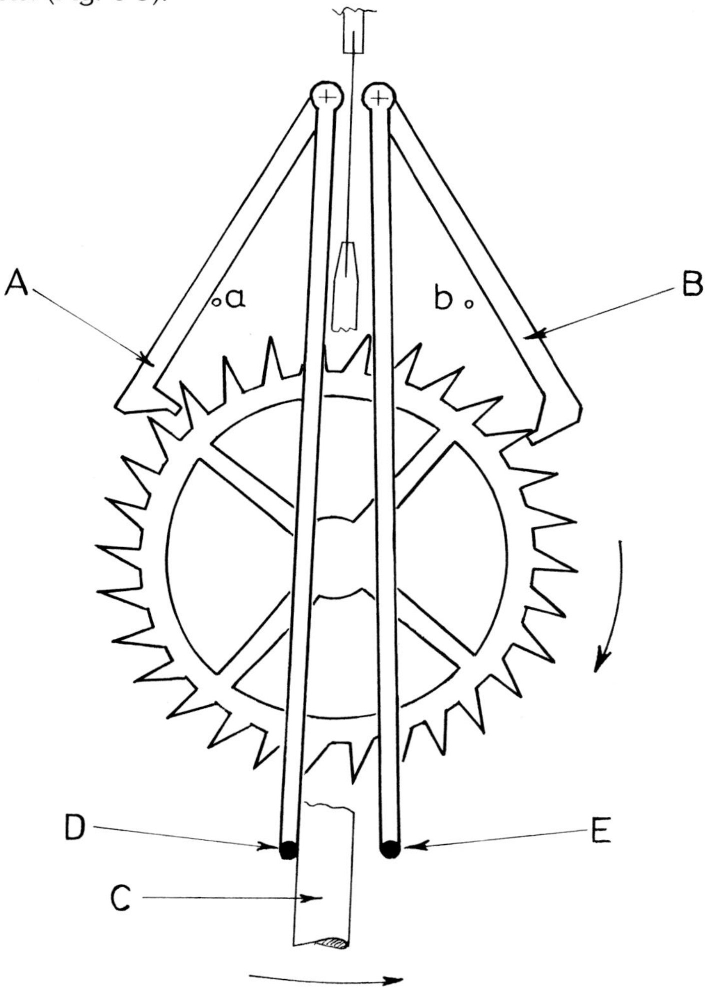

Both Cummings and Mudge's designs can be loosely described as detached gravity escapements as the only work required of the pendulum is the tooth unlocking. However, they suffered from certain deficiencies; Cumming's in that the point of unlocking occurs towards the end of the pendulum's swing, and Mudge's, because the combined function of lifting and locking requires equal forces, the length of the nibs becomes a very critical factor.

A variation of both Cumming's and Mudge's designs is to be seen in an escapement described by W. Nicholson, and made in 1798. (Figs. 5-43, Fig. 6-10). Nicholson claimed, for this escapement, that no lubrication was required. Despite the fact that much of the lifting of the pallets is accomplished by the pendulum itself, there is still sliding action of the wheel teeth on the flat faces of the pallets, which might indicate that lubrication would be beneficial. This one, unlike those of Cumming and Mudge, is not a true gravity escapement as the pallet arms are lifted by the pendulum with assistance from the escape wheel. However, we will return to development of the gravity escapement later. We must now move in another direction.

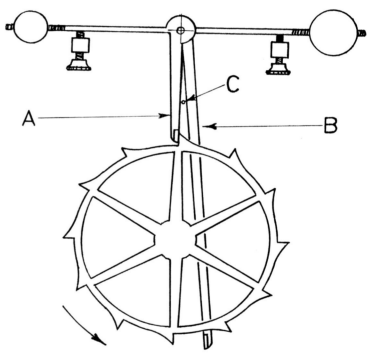

Fig. 6-9. **Mudge's Gravity Escapement.** The escapement consists of two freely pivoted pallet arms **A** and **B**. Firmly fixed to each on its own axis is a long gravity arm which acts upon the pendulum **C** through pins **D** and **E**. There are also banking pins **a** and **b**. The impulse faces of the pallets terminate in small nibs and the escape wheel is shown locked on the nib of the exit pallet **B**.

In the drawing the pendulum is moving to the right and, when it reaches the vertical a wheel tooth will meet with the impulse face of pallet **A**, the pendulum will contact pin E and, as it moves further to the right will unlock the tooth from the nib on pallet **B**. As the wheel turns, pallet **A** will receive impulse and then the tooth will lock on its own nib.

The impulse is given by the weight of the pallet arms acting on the pendulum, but, since the latter is called upon to effect the unlocking, it is necessary to keep the locking faces of the nibs as short as possible. This, in itself increases the possibility, with wear, of the escapement tripping. This requires the driving weight to be kept to a minimum necessitating a high quality of workmanship in the train, and, lubrication of the pallets is still required.

Fig. 6-10. **Nicholson's Escapement.** In 1784 a London clockmaker named Nicholson devised an escapement which he claimed needed no lubrication of the pallets.

The two pallet arms **A** and **B** are pivoted on the same center and each carries its own adjustable weight on an arm resting on stop screws. Each arm has polished agate pallets and the escape wheel is of steel. The pendulum (not shown here) carries a pin **C** which projects between the pallet arms. If we set the clock in motion by moving the pendulum to the left, the pendulum and pin C lift the shorter of the pallet arms with the aid of power from the wheel tooth until the latter escapes, when the wheel will be stopped by the tooth at the bottom coming into contact with the face of the lower pallet and the action repeats to the right.

The principal difference between the action of this escapement and that of Mudge is that, in the case of the latter, the pallet arm is lifted slightly ahead of the pendulum by the sole action of the wheel tooth on the pallet, whereas here, the arm is lifted by the combined force of pendulum and escape tooth. It achieves impulse by receiving the full force of the weight on the downward stroke. Because of the lack of positive locking it would obviously require fine adjustment of the weights to ensure that the force exerted by the train on the escape tooth is not greater than that applied by the weight. In fact, the compound forces involved make the action of this escapement suspect.

The concentrated activity during the latter years of the eighteenth century, initiated principally by John Harrison, to develop a satisfactory marine chronometer, led to experimentation by clockmakers, both in Britain and France, in the use of spring pallet escapements. Edward Massey, a watchmaker working in Hanley in Staffordshire, was awarded twenty guineas in 1803 for two escapements, one of which being partially detached, pointed the way in this direction. He later worked in London and achieved some distinction from the originality of his work and various escapement designs. Fig. 6-11 shows the first of these. Rees' *Cyclopaedia*[10], published in 1820, describes Massey's escapement as ...*of a novel construction, borrowed, probably, from the spring detent escapement of the modern Chronometers and adapted to a clock where the pendulum is free during a large portion of its vibration.*

Fig. 6-11. **Edward Massey's Spring Pallet Escapement.** This escapement is mounted on the backplate of the clock with the escape wheel brought through and supported in a cock **A**. The pallets, **B** and **C**, are in the form of Mudge with impulse faces and locking nibs. They are integral with their arms which turn outwards at their pivot points which are co-axial. The upper part of these arms are sprung apart with a light spring **D** and rest on adjusting screws.

Below their pivot points, the pallet arms are slimmed down to form springs and, planted in the pallet flanks are pins **a** and **b**.

The pendulum carries a cross bar **E**, and each end has an adjustable anvil.

The escapement is shown locked on pallet **C** with the pin **a** in contact with the left hand anvil and giving impulse from the stored energy in the spring pallet arm. This pallet is arrested by the impulse face coming into contact with the next escape tooth and the pendulum continues its arc, free of the escapement, to the right until, shortly before the end of its natural travel the right hand anvil contacts the pin **b** and unlocks the escape wheel. This then turns whilst the tooth on the left lifts the pallet **B** and locks on its nib.

This escapement, unlike Nicholson's claim for his, requires lubrication.

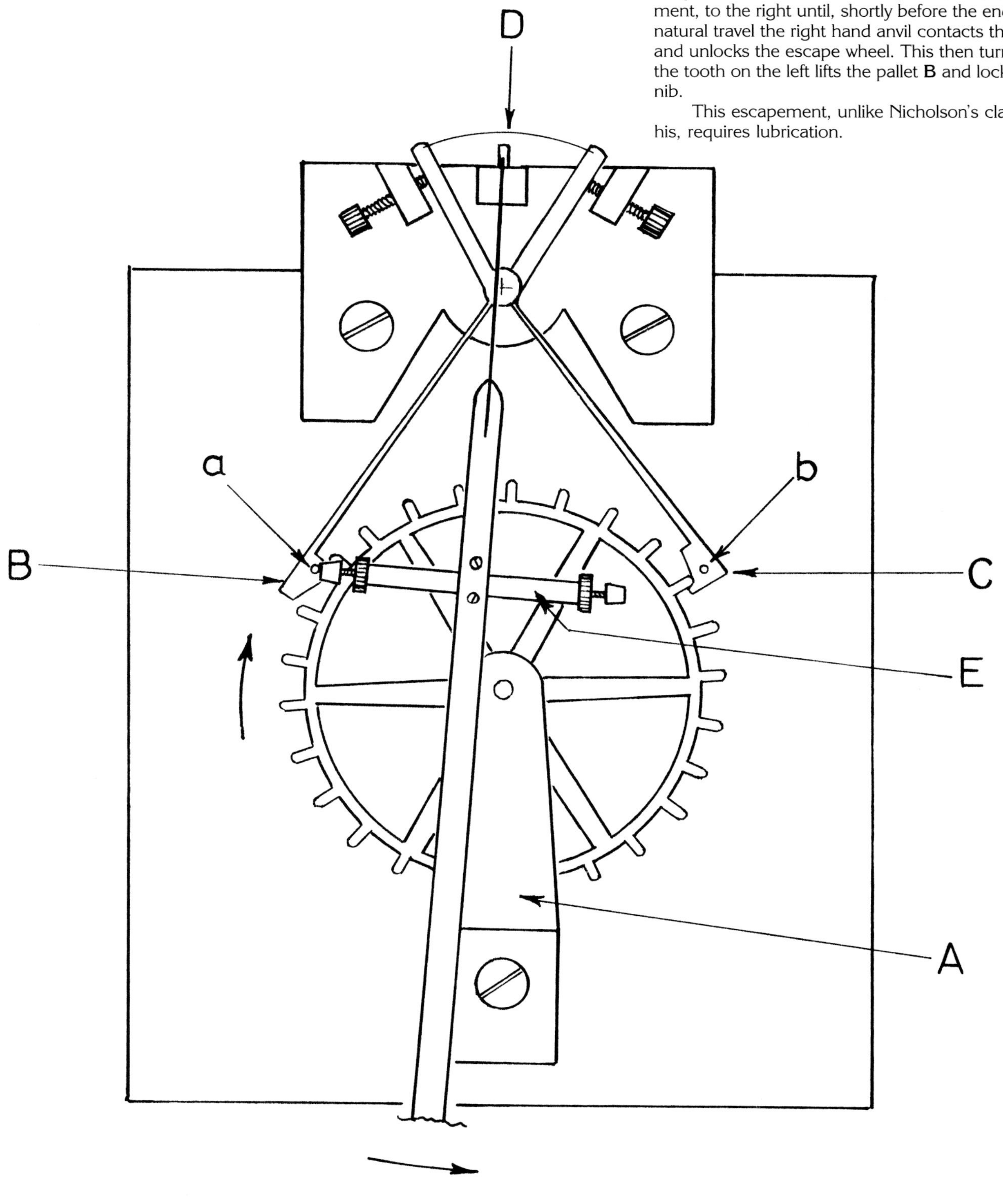

123

However, a celebrated Edinburgh clockmaker, named Thomas Reid, published his *Treatise on clock and watchmaking*[11] in 1826, and in it he describes and illustrates a regulator made for Lord Gray for his observatory at Kinfauns Castle, by his partnership Reid & Auld. Reid's work is fully described in Chapter 16, and his escapement is illustrated in Fig. 6-12. Reid makes no claim to have invented his spring pallet escapement and also makes no mention of Massey, but it does seem likely that details of Massey's design may have reached him.

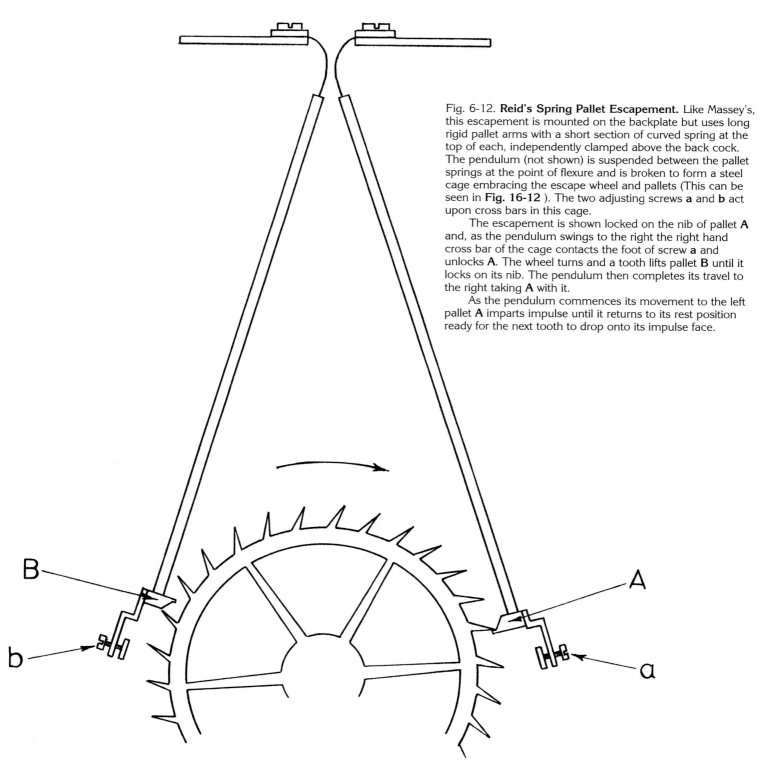

Fig. 6-12. **Reid's Spring Pallet Escapement.** Like Massey's, this escapement is mounted on the backplate but uses long rigid pallet arms with a short section of curved spring at the top of each, independently clamped above the back cock. The pendulum (not shown) is suspended between the pallet springs at the point of flexure and is broken to form a steel cage embracing the escape wheel and pallets (This can be seen in **Fig. 16-12**). The two adjusting screws **a** and **b** act upon cross bars in this cage.

The escapement is shown locked on the nib of pallet **A** and, as the pendulum swings to the right the right hand cross bar of the cage contacts the foot of screw **a** and unlocks **A**. The wheel turns and a tooth lifts pallet **B** until it locks on its nib. The pendulum then completes its travel to the right taking **A** with it.

As the pendulum commences its movement to the left pallet **A** imparts impulse until it returns to its rest position ready for the next tooth to drop onto its impulse face.

A further development of the spring pallet escapement was the design described to the Society of Arts in London in 1820 by William Hardy. A London chronometer and watchmaker, Hardy was awarded fifty guineas and a gold medal, and a succession of orders was received for Observatory regulators. Hardy's work is fully dealt with in Chapter 16 and is shown in Fig. 6-13.

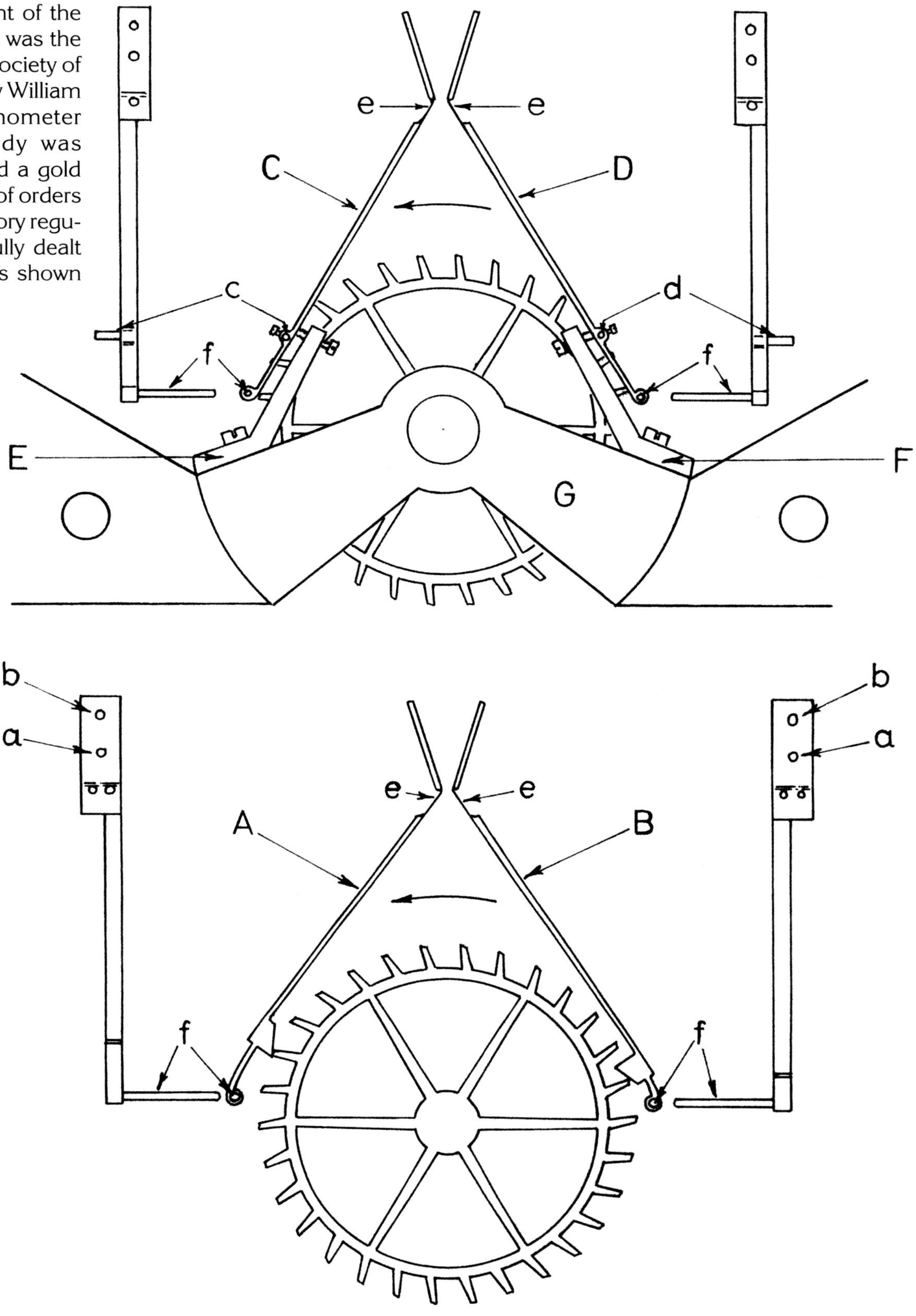

Fig. 6-13. **Hardy's Spring Pallet Escapement.** The details shown here have been re-drawn from Hardy's original submission to the Society of Arts. The escapement, like Reid's, is mounted on the backplate but uses separate spring pallet arms and spring locking detents. The lower illustration shows the escape wheel and the two pallet arms **A** and **B**. These are also shown in elevation at each side. These arms are screwed firmly through the holes **a** to a vee shaped block mounted on the backplate. The holes **b** are tapped to accept short jacking screws to enable the pallet depths to be adjusted.

The upper view again shows the escape wheel, this time embraced by the two locking detent arms **C** and **D**. The detents themselves **c** and **d** are jeweled and the arms rest on brackets **E** and **F** mounted on the backcock **G**.

All four arms are similarly made and mounted and are cut from solid steel with the short section **e** on each reduced in thickness to form the spring on which the action depends. At the foot of each arm is a brass pin **f**. These pins bear on a cross piece fitted to the pendulum (not shown) and, in the case of the detent arms, are used to unlock the escapement. The pallet arms having been raised by the action of the pallets themselves, use their pins to impart impulse.

In France, by the end of the eighteenth century, a number of ingenious clockmakers were producing designs for constant force escapements, many of which were applied to spring driven table regulators, and were described in *Revue Chronometrique*. These were all of the "single beat" type in which the pendulum is given impulse in one direction only. Whilst they all differed from one another in detail, the principles, based on the chronometer escapement, and using either light springs or gravity, remain the same. Three are described here in Figs 6-14A, B, C.

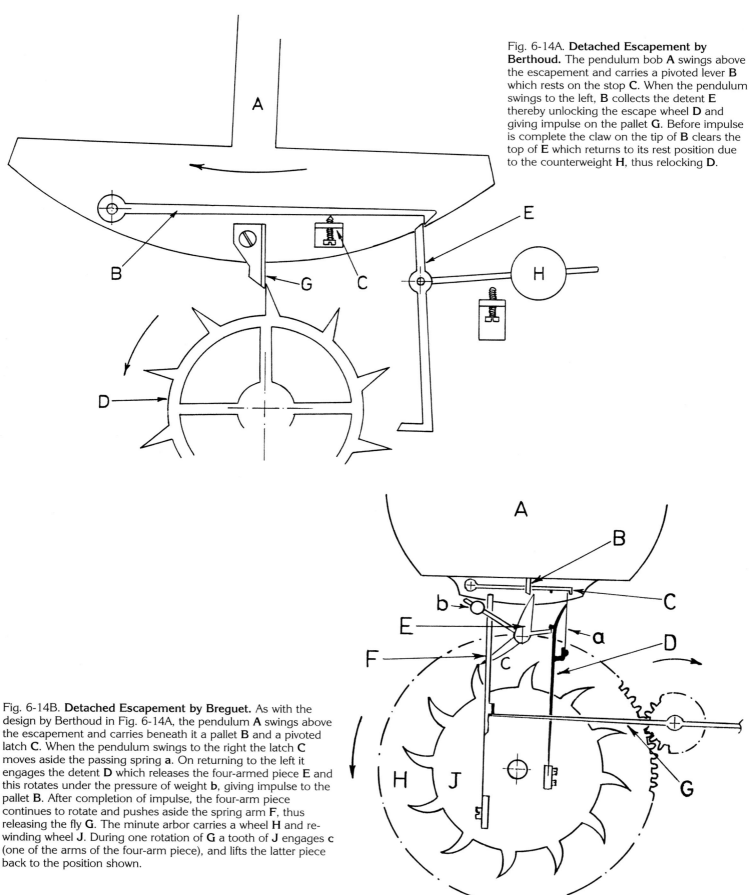

Fig. 6-14A. **Detached Escapement by Berthoud.** The pendulum bob **A** swings above the escapement and carries a pivoted lever **B** which rests on the stop **C**. When the pendulum swings to the left, **B** collects the detent **E** thereby unlocking the escape wheel **D** and giving impulse on the pallet **G**. Before impulse is complete the claw on the tip of **B** clears the top of **E** which returns to its rest position due to the counterweight **H**, thus relocking **D**.

Fig. 6-14B. **Detached Escapement by Breguet.** As with the design by Berthoud in Fig. 6-14A, the pendulum **A** swings above the escapement and carries beneath it a pallet **B** and a pivoted latch **C**. When the pendulum swings to the right the latch **C** moves aside the passing spring **a**. On returning to the left it engages the detent **D** which releases the four-armed piece **E** and this rotates under the pressure of weight **b**, giving impulse to the pallet **B**. After completion of impulse, the four-arm piece continues to rotate and pushes aside the spring arm **F**, thus releasing the fly **G**. The minute arbor carries a wheel **H** and re-winding wheel **J**. During one rotation of **G** a tooth of **J** engages **c** (one of the arms of the four-arm piece), and lifts the latter piece back to the position shown.

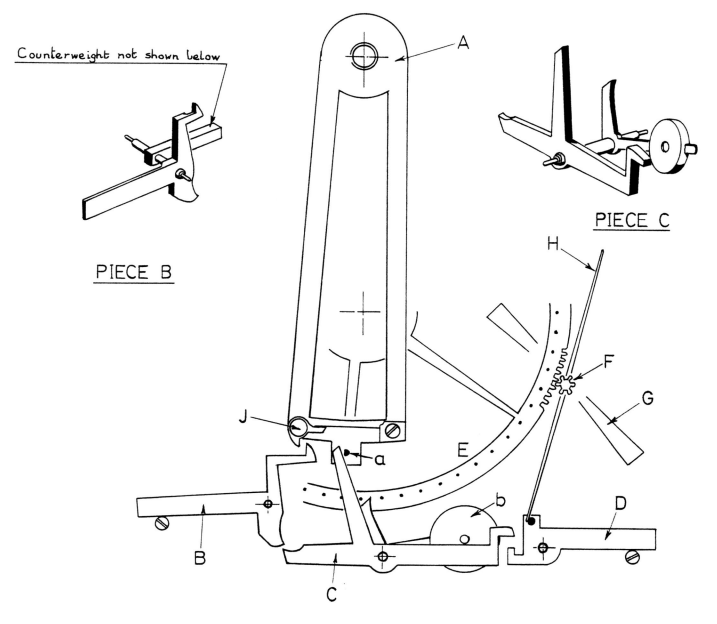

Fig. 6-14C. **Detached Escapement mounted on the front of the dial, possibly by Paul Garnier.** The diagram shows the escapement with pendulum stopped in the vertical position. The sector shaped piece **A** is mounted friction-tight on the arbor carrying the crutch. The pieces **B**, **C** and **D** are all freely pivoted and weighted such that **B** falls to the left on to its stop pin; **D** likewise to the right whilst **C** carries, on its arbor, an adjustable brass weight **b**, causing the piece to be weighted to the right.

The center (seconds) arbor carries a 180 tooth wheel **E** driven from the second train wheel through a 10 leaf pinion. The wheel drives a 6 leaf pinion **F** on which is mounted a steel fly **G** free to move against light friction. Rigidly mounted on the same pinion is a very light double ended steel flail **H**. On the face of the rim of the wheel **E** are set 60 steel pins.

In the stationary position shown, the fly pinion is held by the tip of **H** engaging a pin projecting behind **D**.

If the pendulum is moved to the left the sector **A** also moves to the left and through a pin **a** at its foot causes the piece **C** to rotate anticlockwise on its pivot. If the pendulum is now released it passes through its center point and the left hand arm of piece **C** comes up to the foot of piece **B** and is held back. Screwed horizontally to the bottom bar of the sector **A** is a latch **J** which is integral with a very light passing spring. As the sector continues to the right this latch contacts the dogleg hook at the top of **B** causing the piece to pivot clockwise and unlock **C** which, in turn, also pivots clockwise due to weight **b** on its inner arm and strikes a sharp blow to the tiny anvil on the left hand end of piece **D** which then executes an anti-clockwise jump releasing the tip of the flail **H**.

At this point the pendulum is nearing the end of its travel to the right and it receives a tiny impulse from the vertical arm of **C** falling on to the pin at the bottom of **A**. (It is possible, perhaps, to assume that this small impulse would equalize the loss incurred in the unlatching of the piece **B**.)

When the flail **H** is unlocked the train runs and the flail turns anticlockwise, the tip contacting the side of the sector **A** which has just started to move to the left. The flail tip runs down this side providing the impulse and stopping the train by re-locking behind **D**.

As the train runs, the piece **C** is re-cocked by the inner vertical arm being driven to the left by one of the pins on the face of the wheel **E**.

In 1841, the Astronomer Royal, G.B. Airy, wrote to Edward John Dent proposing a design for a remontoire type escapement based on two dead beat escapements mounted back to back. This development is fully described by Vaudrey Mercer[12] and is illustrated here in Fig. 6-15A.

Dent used the design in a regulator and it employed two similar sized escape wheels. However a variation of the concept is to be found in Chamberlain[13], and is shown in Fig. 6-15B as having a small impulse wheel and large locking wheel. The idea of locking on a larger diameter reduced the power required to unlock the escapement, and this can be seen in a duplex escapement, used many years earlier in a longcase clock by Delander as part of a series of similar clocks that he made. Fig. 6-16.

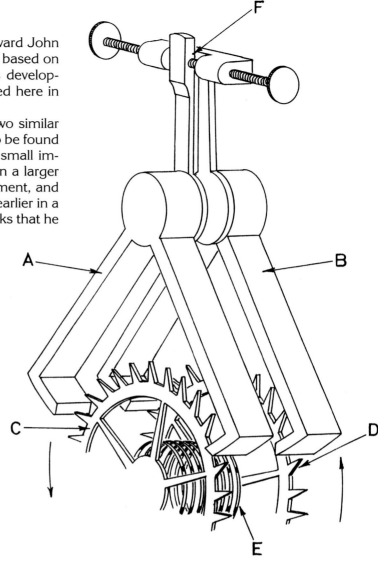

Fig. 6-15A. **Remontoire Escapement by G.B. Airy**. This is one of Airy's escapements as used by Dent. The two anchors **A** and **B** are mounted on a common arbor, **A** being fixed and **B**, which has its pallets reversed, free. The pallets of anchor **A** are conventional and are engaged by the teeth of escape wheel **C** which, in its turn, is freely mounted on the escape arbor. Wheel **D** is fixed to the escape arbor and the two wheels are connected to each other through the light coil spring **E**.

In setting the escapement up, the escape wheel **C** is moved back several teeth, thus tensioning the spring **E**. The two anchors are then adjusted by the mechanism **F** to operate together with anchor **B** locking the fixed wheel **D** and anchor **A** receiving impulse from the free wheel **C**.

The whole escapement is mounted on the backplate with anchor **B** innermost. The pallet arbor carries a conventional crutch with pendulum beat adjustment at its foot.

In this escapement, the pendulum is called upon only to unlock the teeth of **D**, whilst impulse is given by the wheel **C** through the medium of the pre-tensioned spring.

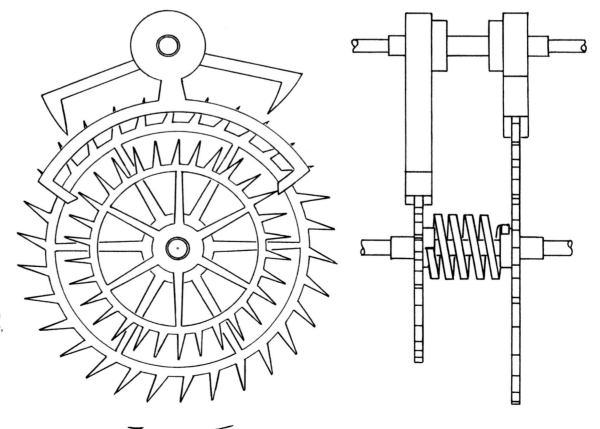

Fig. 6-15B. **Remontoire Escapement by Airy as described by Chamberlain.** This concept of Airy's escapement does seem to be a very good one as, by impulsing through the small wheel and locking on the large one, the force required to unlock the escapement will be less than in the Dent version and, therefore, pendulum disturbance will be reduced.

Fig. 6-16. **Duplex escapement by Daniel Delander.** Although not technically part of the development being discussed in this chapter, it is interesting to take a glance at one of a series of longcase clocks made by Delander early in the 18th century. Here he demonstrates an early appreciation of the need to reduce unlocking friction by mounting two escape wheels on a common arbor. The smaller wheel, between the plates, engages a single pallet and provides impulse whilst the larger, outside the backplate, provides the locking on a short detent. An early example of a single beat escapement.

The next developments that we must examine all involve attempts to reduce the friction loss involved when the pendulum is required to unlock the escapement.

Bloxham's escapement was described in 1853 in the *Memoirs of the Royal Astronomical Society*, and is a development of the Mudge gravity design using a small escape wheel, rather like a nine leaf pinion, and around it a much larger nine tooth locking wheel, thus providing the maximum force for impulse and minimum for unlocking (Fig. 6-17A). Because of the very small size of the escape wheel, the design was very susceptible to variations in power through the train, and if it should trip, could run with potentially disastrous consequences. Fig. 6-17B shows the movement of a regulator with Bloxham's escapement.

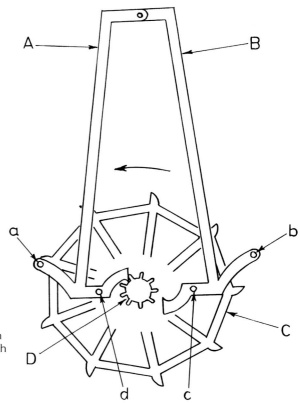

Fig. 6-17A. **Bloxham's Escapement.** Two long pallet arms **A** and **B** are suspended at the same center as the point of pendulum suspension. They hang in front of a nine tooth wheel **C** with the teeth locking on pins projecting to the rear at **a** & **b**.

There is a small wheel **D** in the center similar to a nine leaf pinion which gives impulse. The pins **c** & **d** project forwards and bear on each side of the pendulum.

Fig. 6-17B. **Regulator fitted with Bloxham's Escapement.** This clock is very obviously designed to display the escapement as the movement is, in effect, reversed in the case. Hanging in front of it is the pendulum which is suspended from a heavy frame in the form of a portico.

The time indications are read within the radial cutouts in the plate carrying the escapement. *Photo courtesy of the Science Museum, London.*

It does appear, though, that Bloxham's design was around well before its publication date as that enigmatic and irascible character, E.B. Denison, later Lord Grimthorpe, credits it as the first successful gravity escapement. It certainly influenced Denison's thoughts when, in December 1846, he seems to have had his first ideas on this subject. His gravity escapement then developed directly as a result of his involvement in the design of the Westminster Clock.

Denison was commissioned, together with G.B. Airy, the Astronomer Royal, to design the movement for the Great Clock (Big Ben) in 1851, and the escapement design exercised his mind considerably. In his writings, Denison[14] says that no fewer than five different escapements were tried on the clock before it finally settled down with his double three-legged gravity.

The first three were dead beat escapements and the development of these is described by Denison, culminating in the detached design shown in Fig. 6-18A, B. This escapement was incorporated in a regulator made for Denison by James Brock in 1865, and is described in an article in Antiquarian Horology.[15] It is a development of the last of the dead beat escapements tried in The Great Clock.

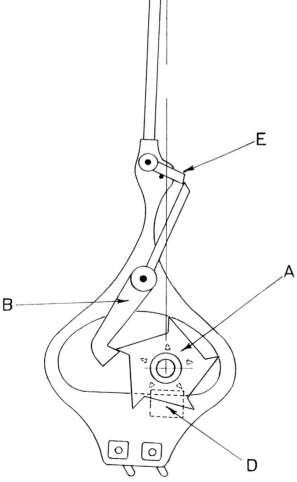

Fig. 6-18A. **Denison's Experimental Detached Escapement.** This single beat escapement was made by Brock and fitted to Denison's own experimental regulator. The action takes place within the pendulum crutch and is, therefore, arranged behind the backplate. The drawing shows the crutch viewed from the backplate. The five pointed starwheel **A** is mounted on the escape arbor and is brought out to the rear to lie just behind the crutch, and is shown locked by the piece **B** which is pivoted on a post fixed to the backplate not, as might appear, on the crutch. The remainder of the assembly is free to swing about the point **C** coincidental with the pendulum suspension point.

The star or escape wheel **A** carries five triangular shaped steel pins that project back through the opening in the bulbous part of the crutch and, fitted to the reverse side of the latter, there is a steel pallet shown dotted at **D**. At the foot of the crutch are two cranked pins which embrace the pendulum rod and are fitted friction tight to provide for beat adjustment.

In the diagram the pendulum is shown commencing its swing to the right. The loosely pivoted latch **E** will immediately cause **B** to rotate clockwise, unlock **A**, and then override **B** as the pendulum continues its swing. **A** will rotate anticlockwise and one of the impulse pins will engage the pallet **D**, whilst, due to it being bottom heavy, **B** will fall back to the right and lock the next corner of **A**.

Fig. 6-18B. **Denison's Experimental Detached Escapement.** This photograph shows the lower part of the crutch where it swings behind the long radial bridge which carries the rear pivot of the escape arbor. The single pallet can be seen screwed to the rear of the crutch and below it the two cranked pins that provide beat adjustment.

Denison then developed his gravity escapement, based on Bloxham's design, the first having a three legged escape wheel, and this was tried in the Westminster clock to be superseded immediately by a four legged design. Denison tells us that the clock was installed in the tower with this escapement and the final change to a double three-legged design made later.

The double three-legged design, with a movement of 60 deg. of the escape wheel on each beat, has a greater facility to cope with the varying load on the hands of large turret clocks caused by wind and snow. We are more concerned with clocks that operate in less taxing conditions and it was Denison's four-legged escapement that was incorporated into many fine regulators from the 1850's on (Fig. 6-19).

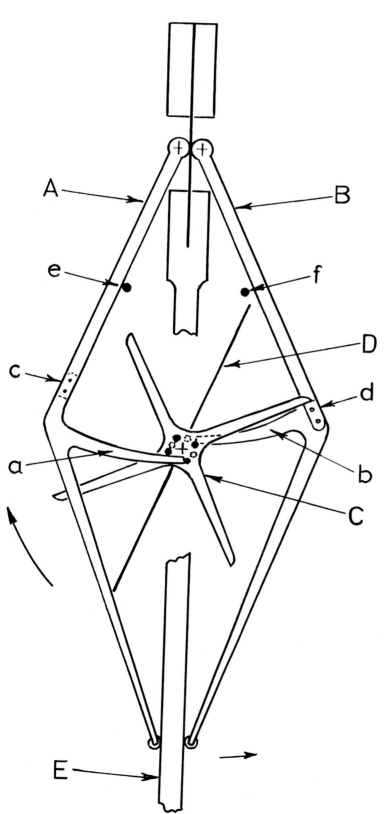

Fig. 6-19. **Denison's Four Legged Gravity Escapement.** Two steel gravity arms **A** and **B** are pivoted independently with **A** in front of and **B** behind the four legged piece **C** which is the locking wheel. Since this piece turns 45 deg. with each beat it is necessary to control its rate of rotation and this is effected by the large fly **D** loosely friction mounted on the escape arbor.

Stops **c** and **d** are fitted to the pallet arms and a set of eight pins, alternately projecting to the front and rear, surround the hub of the four legged piece. These pins engage the ends of the radial lifting pieces that are formed integral with the gravity arms (shown as **a** and **b**) to provide impulse.

At the foot of both **A** and **B** is fitted a roller (usually of ivory) which acts upon the pendulum rod **E**.

Stop pins **e** and **f** prevent the gravity arms from following the pendulum beyond bottom dead center and, in the drawing, the arm **A** is resting on **e** as the pendulum is swinging to the right. Towards the end of this swing to the right the locking arm will escape from the stop **d** and the assembly will revolve causing a pin to press on **a** and lift **A** to the left. On the commencement of this swing to the left, the gravity arm **B** gives impulse to the pendulum through the roller at its foot.

The success of this gravity escapement was undoubtedly due to the long run of the escape wheel with its velocity controlled by a loosely mounted fly, resulting in an escapement which it is almost impossible to trip. A heavy weight is necessary as firstly, with an escape wheel turning once in eight seconds, an extra wheel and pinion are required in the train. In addition, the action of the wheel must be crisp and this has to be obtained against the resistance of the large fly, which requires being something like four inches in diameter. Apart, however, from the extra load on the lower train pivots, the additional weight is largely detached from the pendulum and thus it is more of an aesthetic drawback than anything else.

Despite the unrelenting efforts of clockmakers, scientists and even ingenious barristers for over a century and a half no one had achieved the goal of producing a clock with a truly free pendulum.

It had long been realized that this was an essential prerequisite of perfect timekeeping. The advent of the gravity escapement had removed the principle source of pendulum disturbance, i.e. impulse through a crutch, but the pendulum still required to effect the unlocking and, small though this disturbance may be, it is still significant.

Free Pendulum Escapements

Grimthorpe (Denison) had developed an escapement that was little different from those of Bloxham, Mudge or Cumming in that it delivered its impulse to the pendulum at the extreme part of its swing. This is the point at which it is most sensitive to disturbance and it was realized that the closer to the mid-point of swing the pendulum can be given impulse the better.

This concept is put into practice in the Gravity Impulse, a simple idea that has since been used extensively in electric clocks. This has a pallet that is fixed to the pendulum and is acted upon by a roller mounted on the end of a horizontally pivoted gravity arm, and is then re-set electrically (Fig.6-20). Rawlings[16] quotes Grimthorpe at his most cynical when he describes an earlier version of this idea seen by him in the 1851 Exhibition:

In them the electricity was employed to lift a small gravity arm at every alternate beat, which gave the impulse to the pendulum by falling on a pallet like the down-pallet of a dead escapement, which had the advantage of giving constant impulse when it gave any. But unfortunately it did not always lift. And anyone who sets to work to invent electrical clocks must start with this axiom, that every now and then the electricity will fail to lift anything, however small; and, if his clock does not provide for that it will fail too.

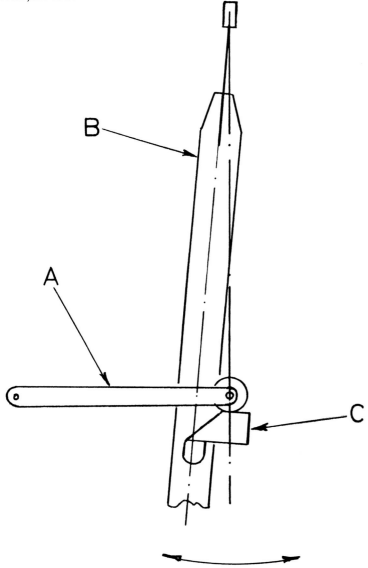

Fig. 6-20. **The Gravity Impulse.** A gravity arm **A**, pivoted on the left, carries a free running roller. The pendulum rod **B** has a single dead beat pallet **C** screwed to it. The pendulum is shown about to swing to the right and the roller lies on the dead face of the pallet. As the pendulum moves, the roller runs down the inclined face of the pallet and gives impulse. As the pendulum commences its return to the left, the gravity arm is re-set electrically, actuated by contacts on the pendulum.

The roller method of giving impulse to the pendulum is dealt with by Denys Vaughan in Chapter 7, but by the mid nineteenth century, the mechanically impulsed pendulum was still required to unlock the escapement and the quest for the free pendulum proved elusive. Of course, a completely free pendulum is an impossibility due to natural effects such as ambient pressure variations, gravity etc. However, it has to be kept in motion by impulses that will in no way affect its natural period of oscillation, and without calling upon it to perform any function other than to keep time. It had long been thought that the first free pendulum clock was made by R.J. Rudd in 1898, and described in the *Horological Journal* in June 1899, and this clock is discussed in the Chapter 7. However, there are at least three claims to pre-date Rudd's work.

The first is described by Cecil Clutton as "Sir William Congreve's Free Pendulum Clock."[17] Nevertheless, Congreve himself did not appear to think the pendulum truly free as, in his Patent, he guardedly refers to it as his "mode of extreme detachment". The pendulum is required to propel a countwheel and, once a minute, unlock the train and, therefore, cannot be considered "free." The second is a series of three regulators made by Bond and Son of Boston, Massachusetts. These clocks used a conical pendulum and are described in Chapter 36.

In 1846, Sir William Thompson, a brilliant Scots mathematician and inventor, became a professor at Glasgow University at the age of twenty-two, and took over the Department of Natural Philosophy there. He was ultimately raised to the peerage in the title of Lord Kelvin, and his work provides us with the third claim.

A communication was published by him in the Proceedings of the Royal Society, 1868-1869, entitled "On a New Astronomical Clock and a Pendulum Governor for Uniform Motion." The communication described Kelvin's first clock that was made for him by James White, a Glasgow optician, with whom he later went into business as Kelvin and White. That business became Kelvin and Hughes Ltd., and around 1966 was acquired by Smiths Industries. Two or possibly three other examples of his clock were made which differ in detail from the original which still stands today where Kelvin originally located it in his University Residence in Glasgow, and another is in the Science Museum.

The story of Kelvin's clock is fully documented in an article by Charles Aked[18], and the principles of its operation are described below. This does genuinely appear to be the first true Free Pendulum clock and also the first to use the principle of a working slave pendulum, and a superior free master pendulum. The later development of this concept is dealt with in Chapter 7, but Kelvin's clock is exceptional in that it is a purely mechanical device.

The essence of the clock is that the movement is controlled by a centrifugal governor. Kelvin had carried out work on the laying of the Atlantic cable, and had designed several types of mechanical governor mechanism to be used in this connection. Seemingly his thoughts turned to the use of a continuously rotating device to drive clock hands as the following extract from his own description would indicate ...*in a clock which I have recently made with an escapement on a new principle, in which the simplicity of the dead-beat escapement of Graham is retained, whilst the great*

Fig. 6-21A

defect, the stopping of the whole train of wheels by pressure of a tooth upon a surface moving with the pendulum is remedied. Fig. 6-21A shows Kelvin's original clock standing in his house at Glasgow University. The basic clock itself is contained in a glazed cast iron case mounted on an iron stand, the whole standing some seven feet in height. The movement is similar to that of a flat bed turret clock and the one hundred and twenty pound driving weight falls some fifteen feet, and is wound through reduction gearing using a crank handle. The greatwheel drives first and second wheels, the third being a contrate wheel. This converts to a vertical drive, and on the driven pinion are mounted a pair of large wheels, the top wheel being of slightly larger diameter as it carries a few more teeth than the lower. These two wheels mesh with two smaller wheels mounted on the lower end of the governor arbor (Fig. 6-21B).

Fig. 6-21A. **Lord Kelvin's Original Clock at Glasgow University.** The "dog-leg" pendulums were Kelvin's own design and proved to be rather unsatisfactory. Each has a cylindrical bob that has a short zinc tube projecting from its axis, and coupled to each is another zinc tube, angled upwards toward the suspension point. The bob is hung directly from the suspension on a platinum wire and, by this means, Kelvin hoped to effect compensation. This appears to be the only clock to have this somewhat bizarre design as he provided glass and mercury pendulums for the later clocks.

The centrifugal governor escapement can be seen on the right below the half seconds' pendulum, and on the left, the free seconds' pendulum, swinging at right angles to the other, and below it the tiny seconds' dial. *Photo by kind permission of Mr. James Aked.*

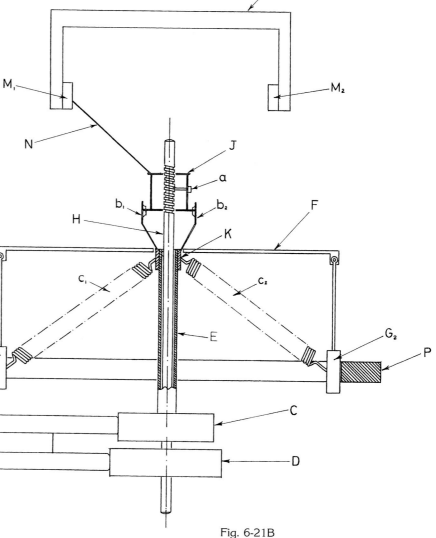

Fig. 6-21B

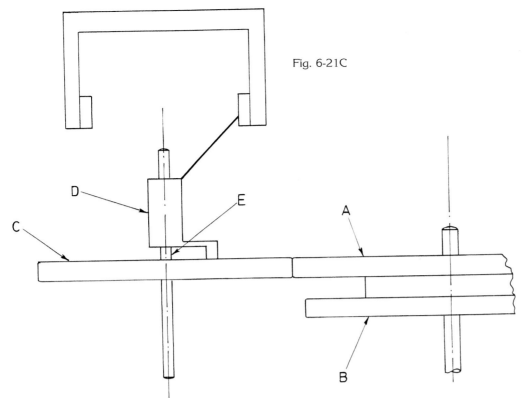

Fig. 6-21C

Fig. 6-21C. Diagrammatic Illustration of Kelvin's Free Pendulum Seconds' Escapement. As in Fig. **6-21B**, **A** and **B** are the wheels that terminate the clock train. The rate of the whole movement having been corrected by the half seconds' pendulum, the wheel **A** is used to drive the free pendulum escapement in a plane 90 deg. from that of the half seconds' pendulum in order to minimize interference.

The free pendulum bob carries the inverted stirrup similar to that in **Fig. 6-21B** with its agate pallets. These are engaged by the wire escapement tooth that is fixed to a sleeve **D** which is loosely mounted on the shaft **E** and driven by a friction pick-up on the face of the wheel **C**.

This wheel **C** is arranged to have a gaining rate of 1/4 per cent. On meeting the pallet the tooth wire gives impulse, but is not released until the exact end of each second. Charles Aked reports that the engagement and release of this tooth is so rapid that it cannot be perceived by the naked eye.

Opposite page :
Fig. 6-21B. **Diagrammatic Illustration of Kelvin's Half Seconds' Escapement.** The clock train terminates in a pair of wheels **A** and **B** screwed together. Wheel **A** meshes with **C** which carries a sleeve **E** to which is fixed a beam **F** (method of fixing not shown). This beam carries a pair of governor weights G_1 and G_2 freely suspended from its ends.

Wheel **D** is integral with an inner spindle **H** which has a left hand square thread near the top. Engaging this thread is a pin **a** which is fitted in a cage **J**. Two leaf springs b_1 and b_2 are fitted to a collar **K** to which are also attached the upper ends of a pair of weak helical springs c_1 and c_2, the lower ends of which are attached to the governor weights G_1 and G_2.

The bottom of the cage **J** is a round flat plate and the ends of the springs b_1 and b_2 embrace this plate and provide a friction lock between **J** and the collar **K**.

The pendulum has, fitted below its bob, an inverted stirrup **L** which carries a pair of agate dead beat pallets M_1 and M_2 These are engaged by a single wire "escapement tooth" **N** firmly fixed to the cage **J**. To complete the layout, the governor weights G_1 and G_2 run within a fixed brass ring **P**.

The gearing of **A** to **C** and **B** to **D** is arranged such that the speed of **C** is 4% higher than that of **D** and the governor is adjusted so that the whole assembly rotates at slightly more than one revolution per second.

The contact of the tooth wire **N** with the pallets takes place very close to the center of the pendulum arc, thus keeping pendulum disturbance to a minimum.

As the tip of the tooth wire climbs the impulse face of a pallet it is momentarily arrested. At this point, due to the differential speed between the inner spindle and the outer sleeve, the assemblies **J** and **K**, held together by friction, will have ridden up the thread slightly and this will have increased the tension in the springs C_1 and C_2 in turn reducing the braking friction of the governor weights against the brass ring.

The whole mass of the governor assembly now speeds up and will continue to revolve at increasing speed until **N** engages the next pallet, is arrested and the pin screws **J** down, thus reducing the tension in springs c_1 and c_2, and increasing the braking effect of the governor weights.

It will be seen, therefore, that this shuttling in speed between impulses is maintained solely by the natural period of the half second pendulum.

The half-seconds' pendulum controls the basic rate of the clock, and the drive to the main dial is taken directly from the movement itself. Kelvin, quite reasonably, decided that the accuracy of the minute and hour display is immaterial in a precision pendulum clock and the only thing that matters is the seconds' indication. Therefore, having obtained the basic timekeeping with his half-seconds' pendulum, he arranged another pendulum beating seconds to control the display of the seconds' dial. This pendulum is suspended under the top of the case, and is arranged to swing in a plane at right angles to the small one so as to minimize any interference between the two (Fig. 6-21C).

If Rudd's clock, some thirty years later, was considered to have a free pendulum then it could be said that Kelvin's clock had two! However, the disturbance of the half-seconds' pendulum is very much greater than the superior one, and to this must be added the force necessary to restrain the escapement tooth, and the necessity for a wide arc to maintain the action, which will increase circular error.

The arc, however, of the superior free pendulum is kept to 0.6deg. This keeps circular error to an absolute minimum, and Lord Kelvin, in his own account of his clock, says that the whole sequence of engagement, retardation, impulse and release is kept within 1/300 second.

Aked says, and I quote: *"ignoring the fact that Kelvin did not place his free pendulum in a partial vacuum, and hence the losses of energy were higher than need be, it may fairly be said that Kelvin's free pendulum was far better than that of W. Hamilton Shortt in the matter of undue interference from the maintaining impulses."*

It seems strange that the significance of Kelvin's work in this area has remained virtually unknown. A debt of gratitude is owed to the late Charles Aked for the scholarly and detailed way he researched his article and placed Lord Kelvin in his rightful position in the horological world as the inventor of the Free Pendulum. Or nearly free? After all, impulse was still required however minimal, and this necessitated contact with the single escapement tooth. But in a train that rotates continuously, the need for the pendulum to unlock the escapement had been eliminated. An extremely lucid explanation of the workings of this extraordinary clock is to be found in Woodward's *My Own Right Time*.[19]

It might be thought that, because the pendulum has to be kept in motion, interference of some kind cannot be avoided. To the purist this is an argument that persists to this day, but let us look at the final phase of the development of the mechanical escapement.

The Last Commercial Mechanical Escapements

Riefler

In 1889 Sigmund Riefler took out a patent for an escapement with an apparently totally free pendulum (Fig.6-22A).

The pendulum suspension spring is firmly mounted in a rigid steel block which has the anchor screwed to its front face. Under the block are two hardened steel knife-edges, and these rest in agate vee pads on the flat face of a massive steel back-cock. There are two escape wheels, that nearest the anchor being for impulse and the other, of slightly larger diameter, for locking. The two cylindrical sapphire pallets have their diameters halved for that part of their length that engages the locking wheel so that the teeth lock on the flat surface. The impulse is given on the circular section.

The pendulum has no other connection with the movement than the suspension itself and the means by which the power is transmitted from the rocking pallet assembly to the pendulum is dependent on the points of the knife-edge being exactly in line with the point of flexure of the suspension (Fig.6-22B).

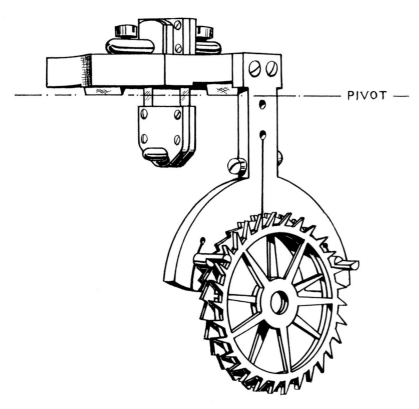

Fig. 6-22A. **Riefler Escapement.** The pallet anchor is securely attached to a substantial cross-piece by two screws. Under this cross-piece are fitted two steel knife edges which allow the whole assembly to pivot on agate pads (not shown) mounted on a massive back cock. Fixed firmly into the center of the rocking piece is the pendulum suspension with its point of flexure exactly on the pivot line.

It will be seen that there are two escape wheels. That nearest the pallet anchor is shaped to provide impulse and the outer one, which has a slightly larger diameter, the locking. The wheels are screwed together so that they rotate as one entity.

The pallets are similar in form to those of a French Brocot visible escapement, being cylindrical, but with the outer ends half cut away, and they are clamped into the anchor frame and capable of adjustment. The outer cut-away part of each pallet is engaged by the teeth of the outer, or locking wheel, and the inner circular part is acted upon by the lifting teeth of the inner or impulse wheel.

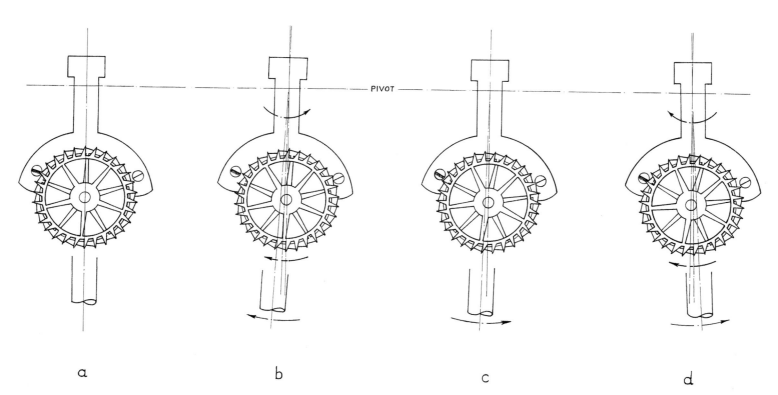

a b c d

Although the action of the pendulum is still required to unlock the escapement, this occurs close to bottom dead center, and minimizes any possible disturbance. The efficiency of this escapement is clearly demonstrated in that the rocking assembly is restricted to one degree of arc, whereas the pendulum takes up a natural arc of three degrees.

Out of general interest, Chamberlain in his book *Its About Time*,[20] in his only reference to Riefler, describes an extremely interesting watch and I quote: *Nearly everyone interested in accurate timekeepers is familiar with the renowned Riefler observatory clocks, but few know that Riefler invented a watch escapement.*

In London, at the house of Charles Frodsham and Company I found a watch of whose history they knew nothing. I secured it as having a very curious escapement and some years later, in going through the files of the Patent Office, found that it was patented December 30, 1890, by Sigismund Riefler of Munich, Germany. The movement undoubtedly was made in Switzerland but bears no number or name. I have never seen another like it nor have any of my friends to whom I have shown or described it. I regret to say that it has disappeared from my collection in some mysterious way.

The balance receives the impulse entirely through the hairspring, which has its outer end pinned to the anchor piece. When the balance is at the end of its excursion it exerts a pull on the pinning of the hairspring, which effects the unlocking, and during the return of the balance it receives impulse from the anchor pallet piece through the outer pinning of the spring. This is, I believe, the nearest approach to a free balance that has been proposed and executed. The interferences to free swing in all the usual forms in receiving the impulse and giving out energy are momentary, but harsh. In this form the impulse is softened through the length of the hairspring as is also the unlocking.

Strasser

In 1899 Professor Ludwig Strasser patented a design for a free pendulum escapement using the same basic principle as Riefler. He put it to use in some of the designs of regulator marketed by his firm of Strasser & Rohde.[21] Whilst the principle is the same, the design differs substantially from Riefler. The pendulum is suspended from a conventional fixed cock, but the lower suspension spring mounting block carries another spring assembly rigidly fixed to it and having its top block over-arching the main suspension (**Fig. 6-23A**).

Bottom (opposing page):
Fig. 6-22B. **The Action of Riefler's Escapement.** In **a** the pendulum is shown at rest. At this point the escape wheel is locked on the left hand pallet. If we now give the pendulum a swing to the left it will, because it is directly attached to the rocking piece through its suspension, carry the anchor with it without bending the spring. This then unlocks the left hand pallet and, almost immediately, the cylindrical part of the right hand pallet comes into contact with the impulse plane of the inner wheel and prevents any further movement of the anchor to the left. Simultaneously, the train being released, the wheels start to turn and, because of the inclination of the lifting tooth, the anchor will immediately start to move to the right leaving the pendulum to complete its free swing to the left. This condition is seen in **b**. This sudden reversal of direction on the part of the rocking assembly against the natural swing of the pendulum causes the suspension spring to bend about its flexure point, thus producing potential energy in the steel.

The right hand pallet, having ridden up the impulse tooth finds itself locked on the projecting tooth of the locking wheel and then the pendulum, having reached the end of its travel to the left, commences its return swing. This condition is shown in **c**. The escapement remains locked as the pendulum continues its swing to the right passing through bottom dead center. At this point the energy conserved in the suspension spring is released as the spring straightens again and this gives the impulse to the pendulum. Continuing its swing to the right, the pendulum comes into line with the stationary rocking assembly and carries it with it to the right. This immediately unlocks the right hand pallet and causes the left hand pallet to contact the impulse plane, arresting the movement of the anchor and starting it back to the left as the escape wheels commence to turn. This condition is illustrated in **d**. The wheels turn until the left hand pallet locks again (as in **a**), whilst the pendulum is free to continue its swing to the right and, in so doing, impart more energy to the suspension spring by bending it to give impulse on its return swing to the left.

It is interesting to note that the operating angle of the rocking assembly in the escapement is only about 1/3 of the arc of swing of the pendulum. The full arc of the pallet anchor is one degree with the pendulum taking up an arc of around three degrees.

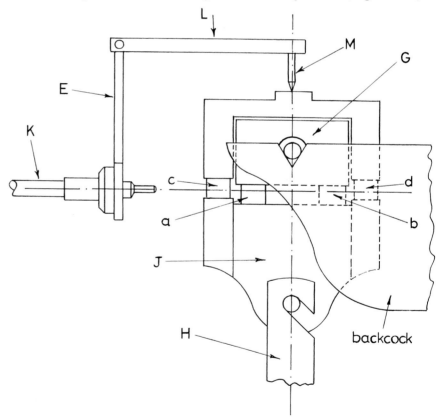

Fig. 6-23A. **Strasser's Escapement–side elevation.** The backcock (mounted separately from the movement) carries the pendulum suspension **G** in a vee notch in the conventional way, and the pendulum **H** is suspended from the lower block **J** on pins whilst **a** and **b** are the suspension springs.

Rigidly attached to the outer edges of **J** are the lower blocks of a further pair of steel springs **c** and **d**. The upper ends fit into a frame which over-arches **G**.

K is the pallet arbor and the vertical piece **E** is shown in **Fig. 6-23B** and carries a beat adjustment and a frame **L** which has a pointer **M** rigidly fitted to it. This pointer acts in a dimple in a pad at the top of the outer suspension frame.

The axis of the pallet arbor is on the line of flexure of the four springs **a**, **b**, **c** and **d** and, although the means by which impulse is achieved is different, the principle remains similar to that of Riefler.

The point of flexure of both sets of suspension springs is carefully arranged to be on the line of the pallet arbor. On the rear of this arbor, where the crutch would conventionally fit, is an arm mounted vertically upwards and carrying a cage assembly that can be adjusted from side to side about the arm with a beat adjustment screw. The outer bar of this cage extends over the top of the suspension and is fitted with a small pointer that locates in a dimple in an adjustable anvil in the top of the upper suspension block.

Unlike the Riefler, this design has a single escape wheel but the anchor is cut to carry separate circular pallets in the manner of a Vienna Regulator. However, each side has two pallet pieces mounted and screwed together, the first to provide the impulse and the second the locking (Fig. 6-23B).

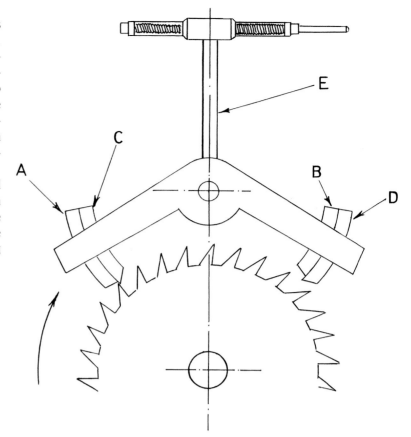

Fig. 6-23B. **Strasser's Escapement – front elevation.** The hardened steel pallets **A** and **B** are in the form of separate radial pieces clamped in slots in the side of the anchor as in a Vienna Regulator, but, in this case each has another similar piece, **C** & **D** clamped to its back. These pieces, though, have a reverse angle at the tip and act as detents to lock the escapement. Unlike the nibs on Mudge pallets they can be adjusted to provide the optimum locking angle.

The vertical arm **E** rises from the rear of the pallet arbor and carries a beat adjustment. (See Fig. 6-23A).

Leroy

The final design of escapement that we should perhaps examine is one used by Leroy et Cie., in the first half of the twentieth century for observatory clocks of their manufacture. The clock itself is extraordinarily simple, consisting of a thirty tooth escape wheel which is driven through a pinion by another wheel which turns once in 7-1/2 minutes. This wheel is driven by a gravity arm lifted every thirty seconds by an electromagnet energized from a four volt battery. The clock has no hour or minute indications.

Fig. 6-24 shows that the pallets are of the form used by Mudge with nibs at the extremities of the impulse faces to provide the locking. They are made of sapphire and are fixed to the end of arms that have a short section near their roots reduced to the thickness of a suspension spring. The centerline of the escape wheel is above the pendulum and attached to the top of the latter is a rigid "Y" frame. This frame carries the two pallet arms, which rest on adjustable stops when the spring is not being tensioned by the action of the escape wheel.

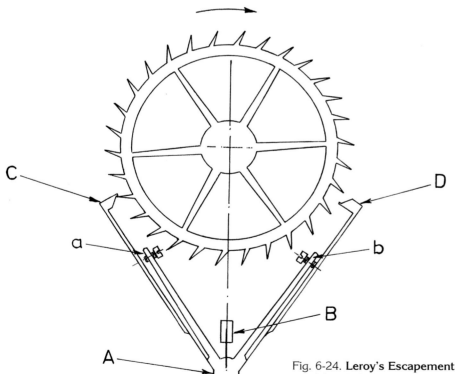

Fig. 6-24. **Leroy's Escapement.** A substantial 'Y' frame **A** is fitted to the top of the pendulum below the suspension **B**. The escape wheel is mounted above the pendulum and each of the sapphire pallets **C** and **D** is fitted to a long steel arm that is reduced in thickness near its root to form a spring.

The pallets each have a locking nib as used by Mudge, and, unless the pallet spring is tensioned by the escape wheel, the arms rest on adjustable screws, **a** and **b**.

The escape wheel is seen locked by the pallet **C** and when the pendulum swings to the right the tooth will unlock, the wheel turn and a tooth will fall onto the impulse face of pallet **D**, tensioning its spring and locking on its nib.

It will be seen that this design is a simplification of that of Reid (See Fig. 6-12).

Once again, the pendulum is required to perform the unlocking. However, the pendulum takes up a total arc in excess of 2 deg., with an escaping arc of 1.4 deg., but the unlocking is kept to a minuscule 0.1 deg., thus minimizing any disturbance.

Rawlings[22] quotes the record of a year's run of Leroy's clock no.1448 at Neuchatel Observatory in 1935 when the difference between its fastest and slowest rates throughout the year was 0.041 of a second per week.

The Mechanical Escapement Becomes Redundant

Looking back over the two hundred fifty years of work that went into the development of the mechanical escapement for pendulum clocks, the central objective was always to free the pendulum from interference whilst still maintaining a means to keep it in motion.

Despite the radical approach of Lord Kelvin who dispensed with the need to unlock the escapement, and the work of Riefler and Strasser, the answer was to lie in the use of electricity, and this subject is dealt with by Denys Vaughan in Chapter 7.

Reference should also be made to the Bond regulator, which was a brilliant concept using a conical pendulum and therefore, like Kelvin's, had a continuously running train. This nineteenth century attempt to overcome the problem is covered by Don Saff in Chapter 36.

Lord Grimthorpe's gloomy predictions about the inability of electrical power to perform the function of re-setting a small gravity arm could be forgotten as Denys Vaughan shows. Even more, Grimthorpe could not possibly have foreseen the electronic revolution that was to consign Huygen's pendulum to the horological history books.

References

[1] Kesteven, M.J.L. *Burgi and the Pendulum Controlled Escapement.* Antiquarian Horology. Winter 1980. p. 441.

[2] Greenwood, C. *A Joseph Knibb Longcase Clock with Early Anchor Escapement.* Antiquarian Horology. Spring 1988. p. 259.

[3] Dawson, Drover and Parkes. *Early English Clocks.* Antique Collectors Club. 1982. p. 132.

[4] Roberts, Derek. *Precision Pendulum Clocks.* Private publication. 1986. p.51.

[5] Howse, D. *The Tompion Clocks at Greenwich and the Dead Beat Escapement.* Antiquarian Horology. December, 1970, and March 1971.

[6] Reid, T. *Treatise on Clock and Watchmaking.* p.192.

[7] Chamberlain, P.M. *Its About Time.* Richard R. Smith, New York. 1941. pp 144, 145 & 146.

[8] Roberts, Derek. *Continental and American Skeleton Clocks.* Schiffer. 1989. p. 124.

[9] Cumming, A. *The Elements of Clock and Watch-work Adapted to Practice in Two Essays.* 1766.

[10] Rees, A. *Rees' Clocks, Watches and Chronometers.* Extracts published by David and Charles 1970. p.211.

[11] Reid, T. *Treatise on Clock and Watchmaking.* p.195.

[12] Mercer, Vaudrey. *Edward John Dent and his Successors.* The Antiquarian Horological Society. 1977. p.121.

[13] Chamberlain, P.M. *Its About Time.* Richard R. Smith, New York. 1941. p.187.

[14] Denison, E.B. *Clocks, Watches and Bells.* Lockwood, Sixth Edition. 1874. p.116.

[15] Martin and Roberts. *Lord Grimthorpe and his Experimental Regulator.* Antiquarian Horology. June, 1983. p.157.

[16] Rawlings, A.L. *The Science of Clocks and Watches.* Pitman. 1944. p.85.

[17] Clutton, C. *Sir William Congreve's Free Pendulum Clock.* Antiquarian Horology. June, 1958.

[18] Aked, C. *The First Free Pendulum Clock.* Antiquarian Horology. March, 1973. p.136.

[19] Woodward, P. *My Own Right Time.* Oxford University Press. 1995.

[20] Chamberlain, P.M. *Its About Time.* Richard R. Smith, New York. 1941.

[21] Kummer, H. *Ludwig Strasser.* Callwey. 1994. p.161.

[22] Rawlings, A.L. *The Science of Clocks & Watches.* Pitman. 1944. p.76.

Chapter 7
Electric Clocks
by Denys Vaughan

The aim of this chapter is to trace the evolution of the precision electric clock by examining the most significant examples of each type. Precision clocks by their very nature have to be accurate but they also have to be reliable. In the rudimentary electric clock these requirements are incompatible, as the firm electrical contact required to reliably switch the current can only be obtained by interfering with the free motion of the pendulum or balance. The development of the precision electric clock is therefore an attempt to reconcile these conflicting demands.

Electric clocks can be conveniently divided into three categories:

Electromagnetic pendulums –in which the pendulum is impulsed directly by an electromagnet.

Electric remontoires –in which the pendulum is impulsed by a gravity arm or a spring, which is reset electromagnetically.

Electrically rewound movements–conventional mechanical clocks which are rewound electromagnetically.

Only the first two categories will be dealt with in here, as the essential features of the clocks in the third category are described elsewhere in this book.

Electromagnetic Pendulums

The first electric clock of this type was patented by Alexander Bain in 1841 although he developed it into a more practical form four years later.[1] Its mode of operation can be seen from Fig. 7-1, which is taken from Bain's *Short History of the Electric Clocks*.[2] A solenoid (B) on the end of the pendulum rod swings freely over two magnets, which are mounted with similar poles facing inwards. The arrows indicate the path that the current takes as it flows from the battery to the solenoid, via the two suspension springs, returning through the switch (abcfg). A plan view of this switch is shown at the bottom of the figure. Half of the disc on the left-hand side is made of gold and the other half of agate (shown darker). A groove is cut across both of them and a gold stud is inserted flush with the surface of the groove in the agate. The right hand disc contains a strip of gold with a groove parallel to the one in the other disc. The gold strip and the gold stud are connected to the wires which enter and leave the switch. The gold half of the left-hand disc is used to drive slave clocks and can be disregarded in the present context. A light bar (gf) rests on gold points in each groove and is moved alternately to the right and left by the action of the pendulum. When it is in the right-hand position it connects the gold stud to the gold strip and energizes the solenoid giving the pendulum an impulse to the left. At the end of its travel to the left the pendulum moves the bar off the stud so that the current stops, allowing the pendulum to return to the right. This moves the bar into contact with the stud again and the process is repeated.

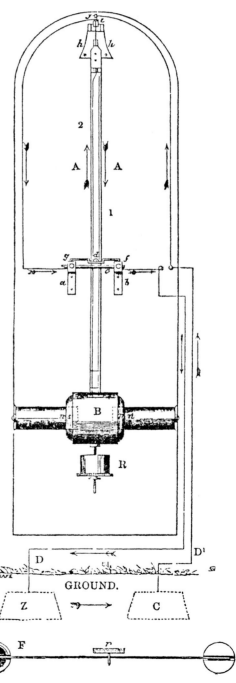

Fig. 7-1. Alexander Bain's electric clock, from *A Short History of the Electric Clocks*.

A defect of this type of clock is that the amplitude of the pendulum varies with the state of the battery and the early batteries were particularly unstable. Bain was aware of this and his clocks had two features that reduced this effect. Firstly the clocks were powered by an earth battery, which Bain had patented in 1843[3] This is shown schematically immediately below the clock in Fig. 7-1 and consists of plates of zinc and copper (or carbon) buried at least three feet deep (1 meter) in the ground and no less than four feet apart (1.3 meters). Because the temperature of the earth is reasonably constant the output of this battery was more stable than that of a contemporary battery of conventional design. Secondly, if the amplitude of the pendulum increases it will move the bar further so that it is only partially on the gold stud and the current in the coil will be reduced and any further increase in amplitude will move it completely off the stud so that the current is cut off. Bain claimed that this would stabilize the amplitude of the pendulum, provided that the battery was sufficiently powerful to always move the bar at least as far as the gold stud. Presumably this encouraged him to construct the regulator with mercury compensation pendulum which is illustrated in his *Short History of the Electric Clocks* and reproduced here as Fig. 7-2. This shows the alternative design in which the solenoid is fixed and the magnet is mounted on the pendulum. Whatever its merits as a timekeeper the Bain clock was notoriously unreliable due to the poor electrical contact made by the light bar and cannot be considered a viable precision clock. The design was however revived by Bentley with some success in 1910 using a more positive switch, which also reversed the current in the coil when the amplitude of the pendulum became excessive.[4] A Bentley regulator with slate back and invar pendulum is preserved at the Science Museum in London (Inv. No. 1982-666). Although no rigorous ratings have been published for the Bentley clock contemporary accounts suggest an accuracy of within one minute a year.[5]

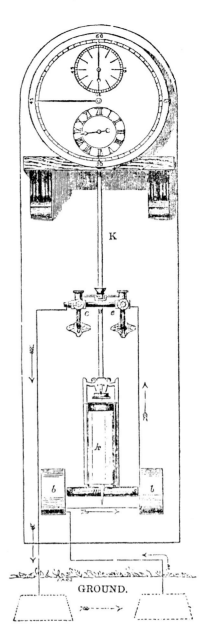

Fig. 7-2. Alexander Bain's regulator clock with mercury compensation pendulum.

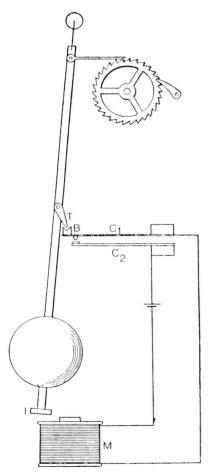

Fig. 7-3. The mechanism of the Hipp clock.

The switching problem, which was the Achilles heel of the Bain clock, was overcome in a completely different way by Matthäus Hipp. In 1843 Hipp described a mechanical means of maintaining an oscillating pendulum by delivering an impulse when its amplitude fell to a certain value. The virtue of this method was that the pendulum could swing freely for quite a long time between impulses. The purely mechanical device does not seem to have been taken up and it only achieved fame when Hipp applied the same principle to operate an electric clock. It is not known precisely when this transition occurred but electric clocks with the Hipp toggle, as it later became known, were produced at the telegraph factory which Hipp founded in Neuchâtel in 1860. The mechanism of the early clock is shown schematically in Fig 7-3. The toggle (T) mounted on the pendulum is brushed aside by the block (B) as the pendulum oscillates. Eventually the amplitude of the pendulum will decrease to such an extent that the toggle will fail to clear the block and come to rest in the notch on its top surface. On the return swing of the pendulum the block will be forced downwards closing the electrical contacts and energising the electromagnet (M). This impulses the pendulum by attracting the soft iron armature (I) at its lower end. The freedom of the pendulum was obviously important from the timekeeping point of view but the long interval between switching was equally important from the reliability point of view as a firmer contact could be made without adversely affecting the pendulum. In 1881 Lemoine of Paris introduced another method of prolonging the interval between switching, although in this instance the pendulum did not swing entirely freely between impulses.[6] Lemoine used a count wheel that was advanced, one tooth at a time, by the pendulum. Electrical contact was established by arms mounted on this wheel, which energized an electromagnet placed below and slightly to one side of the pendulum bob. The count wheel does not appear to have been applied to other electromagnetic pendulums although it was later used to great effect in electrically reset gravity remontoires.

At Neuchâtel, Hipp established a fruitful relationship with the Director of the Observatory, Dr Adolph Hirsch, and during the period 1877 to 1884 he refined his electric clock design to produce a regulator for the Observatory (Fig 7-4A). One of the defects of the basic Hipp clock is that the toggle oscillates after it passes the block and this can apply small random impulses to the pendulum. This was overcome in the "dead beat" mechanism devised for the observatory clock and shown at II in Fig. 7-4B. The block (g) is now mounted on the pendulum and the toggle (r) is inverted. As the block approaches the toggle from the left it will brush it aside lifting the counterweight (t) on the left. After the block has passed the toggle the counter weight will flip it over to the position shown on the right of the figure. To reduce friction and eliminate the need for lubrication switching is done by means of levers mounted on knife-edges and sparking at the electrical contacts is almost entirely eliminated by short circuiting the electromagnet just before the contacts open. The electromagnet attracts an armature on the pendulum (shown at III in Fig. 7-4B) providing a short impulse just before the pendulum has reached its mid-point. Although the minimum amplitude of the pendulum is fixed, the maximum amplitude is still dependant on the output of the battery which must be kept as constant as possible. The swings of the pendulum are counted electrically by means of a switch mounted at the top of the pendulum (shown at I in Fig. 7-4B) which produces pulses of alternating polarity every second. Because the clock did not require winding or lubricating it could be mounted in a glass-sided tank and kept at a constant low pressure for several years without disturbance. In 1884 and again in 1891 Dr Hirsch reported on its performance and showed that the average variation from its daily rate was only 0.03 sec,[7] comparable to the best mechanical clocks of that period.

Fig. 7-4A. Hipp observatory clock, formerly at the Neuchâtel observatory (Musée international d'horlogerie, La Chaux-de-Fonds).

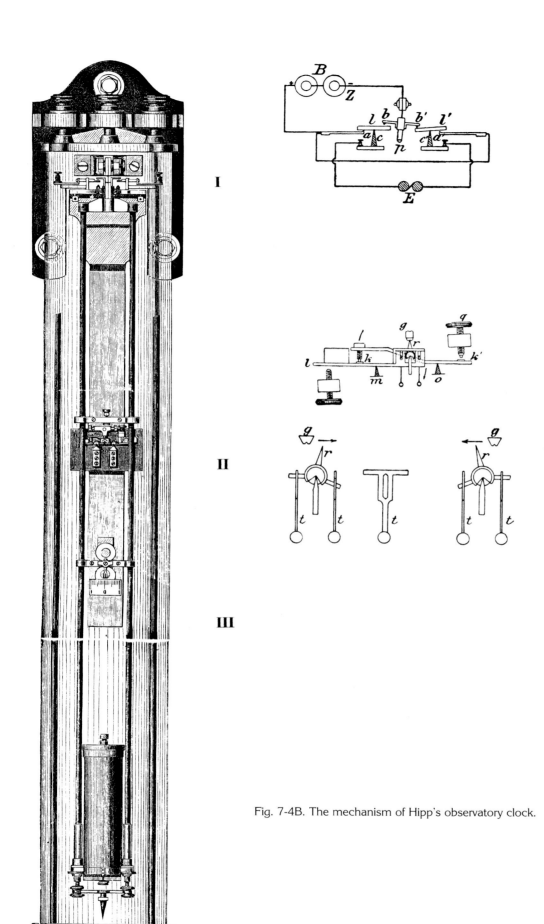

Fig. 7-4B. The mechanism of Hipp's observatory clock.

Clocks of the Hipp type were subsequently made by many different manufacturers, in parallel with improved versions of the generic Bain type. A rigorous mathematical analysis of the operation of this latter type of clock was carried out by Cornu towards the end of nineteenth century [8] and made more widely accessible through the writings of Lavet.[9] The results that emerged from this study were to have a profound effect on the development of the electric clock in France. The switching mechanism was arranged so that it delivered a short impulse at the mid-point of the pendulum's swing, so going some way towards meeting Airy's criterion for minimizing the escapement error. The design of the permanent magnet was also greatly improved so that less power was required to keep the pendulum in motion and there was consequently less wear on the switch. The strong field of the moving permanent magnet induces a voltage in the impulsing solenoid that opposes that of the battery. An increase in the amplitude of the pendulum increases its velocity and as this will increase the induced voltage the impulsing current will drop, so tending to stabilize the amplitude. These features were incorporated in the electric clocks of Professor Féry and later those of the Brillié brothers. A Brillié clock with a seconds pendulum of invar, similar to the one shown in Fig. 7-5, was used at the Paris Observatory to synchronise a series of half-second pendulums of similar design. A schematic diagram of the Brillié clock is shown in Fig. 7-6. As the pendulum moves from right to left, a click attached to it advances a wheel by one tooth. Another tooth on this wheel closes the electrical contact when the pendulum approaches its mid point, energizing the solenoid and giving an impulse to the pendulum. Two soft iron slugs are mounted on either side of the solenoid and provide a means of making fine adjustments to the rate of the clock. Raising the position of the slugs allows them to attract the magnet on the pendulum more strongly, increasing the force that returns the pendulum to its zero position and so increasing its rate. Current consumption was of the order of 1 ma and the clocks were initially driven by constant voltage batteries of the Latimer-Clark type and later by Féry's version of the Leclanché cell, which used air as the depolariser.[10] It was claimed that a Féry cell would run the clock for several years with a daily rate of around 0.1 sec.

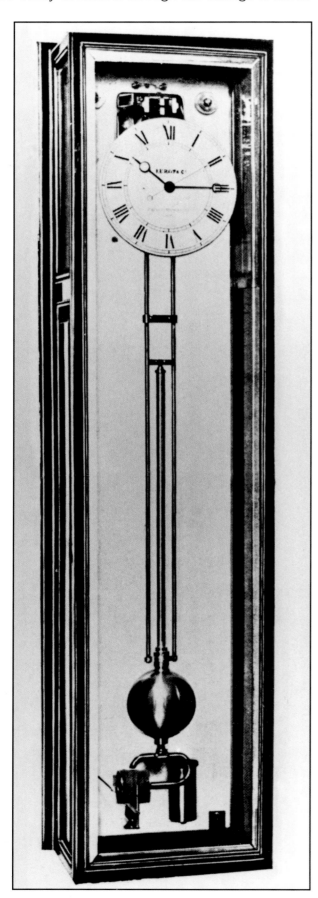

Fig. 7-5. Brillié clock with seconds pendulum.

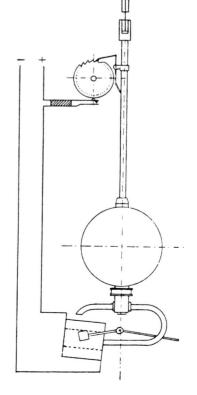

Fig. 7-6. The mechanism of the Brillié clock.

Electronic Switching

In the clocks described so far, the current has been switched by the pendulum closing contacts, which interferes with its free motion. The introduction of the transistor offered the possibility of reducing this interference by eliminating the mechanical contacts and switching the current electronically. It also removed the other defects associated with the mechanical switch, such as dirty contacts and variable contact resistance. The first successful transistor switch was devised by Maurice Lavet and patented by the firm of Leon Hatot in France in 1953 and in Britain in the following year[11], although the principle had been used much earlier by M. H. Abraham using a thermionic valve. It was applied to the ATO clock, which was similar in principle to the Brillié, and its mode of operation can be seen from Fig. 7-7. The voltage induced in the sensing coil (BC) when the permanent magnet attached to the pendulum moves in a particular direction will switch on the transistor amplifier. It then amplifies the current flowing in the sensing coil and feeds it to the impulsing coil (BE). The energy for switching is of course still abstracted from the pendulum, due to the interaction of the sensing coil with the moving magnet, but this requirement can be reduced a hundredfold by the gain of the amplifier. This arrangement was later used by Fedchenko to drive his AChF-3 regulator, which is described in Chapter 37. Fedchenko used ALNICO type magnets to produce a highly concentrated magnetic field which enabled him to give a short sharp impulse to the pendulum as it passed through its centre point (see Figs 37-19, 20). He also used a two-stage amplifier that reduced further the interference with the free motion of the pendulum. Rather surprisingly, having used electronic switching to impulse the pendulum Fedchenko still relied on mechanical contacts to count the swings of the pendulum and operate the slave dials.

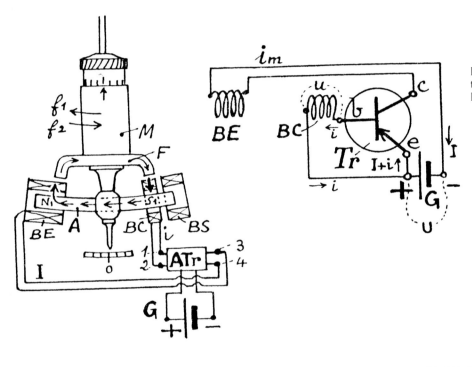

Fig. 7-7. Hatot's method of using a transistor to impulse a clock, from British Patent No. 746,465.

It was obviously desirable to switch the current without abstracting any energy from the pendulum and the discovery of devices which give an electrical response to light suggested a way in which this might be achieved. This was attempted early in the twentieth century by Schlessor, who fitted a vane to the bottom of a pendulum so that it intercepted a beam of light as the pendulum swung through its central position. The light was focussed on a selenium plate, whose resistance was sensitive to light, and this was used to control the current to operate a relay, which switched the current to the impulsing coil. Due to the slow response of the selenium this arrangement was not successful and the method only became viable with the introduction of the photoemissive cell. Fig. 7-8 is a schematic diagram of a clock devised by General Ferrié that was probably the first to use such a photocell for this purpose. A concave mirror (2) mounted on the pendulum reflects a beam of light on to the photocell (3) as the pendulum passes through its mid point. This produces a short pulse of current, which after amplification, is used to impulse the pendulum. This method was used later by Schuler in his observatory clock, which had several other interesting features.[12] Schuler's original arrangement resembled the Shortt free pendulum clock (see pg. 156) in as far as it used a Riefler clock as a slave but this was found to be superfluous and was abandoned in his second version, which was introduced in 1937 (Fig. 7-9A). A vane at the bottom of the pendulum (Fig. 7-9B) blanks off the light to a photocell as the pendulum passes through its zero position. The weak signal from the photocell is amplified by a thermionic valve and used to operate a relay (Fig. 7-9C). Thus when the pendulum swings in one direction, current passes through the impulsing coil and charges a capacitor and on the next swing the capacitor is discharged reversing the flow of current through the coil. The impulsing coil has the shape of the figure "8" and lies between the poles of a horseshoe magnet mounted on the top of the pendulum (Figs 7-9B, 7-9D). When the pendulum is stationary the poles of the magnet are opposite the crossover point of the coils ensuring that the pendulum receives a short impulse at the centre point of its swing in either direction. The voltage of the power supply was stabilised in an attempt to keep these impulses constant but the impulsing current was still dependant on the resistance of the impulsing coils, which could vary with temperature. The pendulum of the Schuler clock (Fig. 7-9B) is unusual as it swings on knife-edges and is of the minimum period type. This has been achieved by loading the pendulum above the knife-edges until the centre of gravity coincides with the radius of gyration. When these conditions are met the period of the pendulum is insensitive to changes in the position of the suspension point, thus minimising the effects of wear on the knife-edges. Although this clock had great potential, its long-term reliability was questionable due to the relatively short life of thermionic valves. The same problem was overcome in early quartz clocks by running them in triplicate, where this extravagance could be justified by their superior performance. The Schuler clock was unable to compete with the emerging quartz clocks and only a very limited number were produced by the firm of Riefler. The use of a beam of light to determine the position of the pendulum became a more practical proposition with the advent of solid state devices and it was applied successfully to the Littlemore clock which is described in Chapter 38.

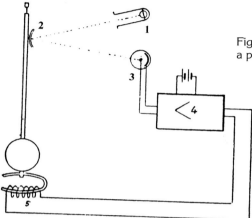

Fig. 7-8. Ferrié's method of using a photocell to impulse a clock.

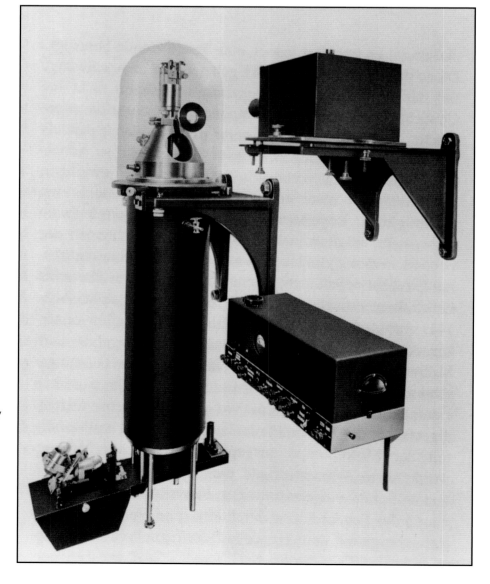

Fig. 7-9A. Schuler observatory clock, as made by Riefler.

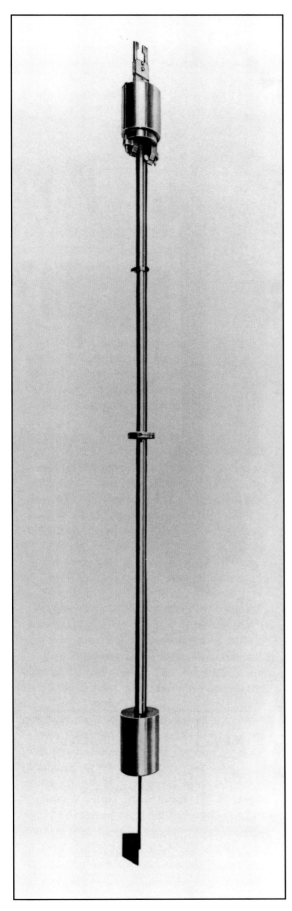

Fig. 7-9B. The pendulum of the Schuler clock.

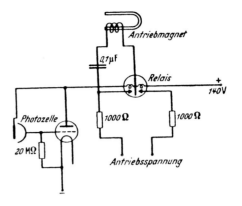

Fig. 7-9C. The method used to impulse the Schuler clock.

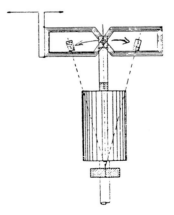

Fig. 7-9D. The impulsing coil and magnet of the Schuler clock.

Electric Remontoires

As we have seen, Alexander Bain was aware of the detrimental effect that the variable output of contemporary batteries could have on the timekeeping of electric clocks and in 1842 he displayed a clock at the Polytechnic Institution in London in which the pendulum was impulsed by a spring which was tensioned by an electromagnet.[2] The spring was retained electromagnetically, so that the current was flowing for half the time, and this high current consumption led him to abandon this design in favor of the alternative methods of controlling the amplitude of the pendulum that have been described earlier. Charles Shepherd overcame the high current consumption by using a detent to hold the spring and his patent of 1849[13] describes both an electrically reset spring remontoire and a gravity remontoire. Although Shepherd expressed a preference for the spring remontoire on the grounds that the friction was variable in the gravity remontoire all the surviving examples of his work used the latter system. His clocks will therefore be dealt with in the section on gravity remontoires.

Spring Remontoires

At the Universal Exhibition in Paris in 1855 the French instrument maker Paul Gustave Froment displayed a simple and elegant electric clock[14] which operates in the following manner (Fig. 7-10). A weight (e) attached to the end of a spring (h) is supported by an arm (j) which is pivoted at (k) and forms part of the armature of the electromagnet. When the pendulum swings to the left, a screw (d) attached to it, collects and raises the spring. This establishes electric contact and energizes the electromagnet (g) which lowers the arm (j). When the pendulum returns to the right the spring and weight will therefore exert a force on the pendulum over a longer distance than the pendulum exerted on the spring during its movement to the left. This difference provides a constant impulse to maintain the oscillation of the pendulum. In this respect it resembles the mechanical gravity escapement but with the advantage that the pendulum is not required to unlock the gravity arm. Electrical contact is made fairly firmly as the pendulum is initially acting against the force of the spring and it can be argued that the energy taken from the pendulum to establish this contact is returned to the pendulum on its return swing. However current consumption is high as the electromagnet is energized for around a third of the time and the interference with the free motion of the pendulum occurs at the end of its swing. Despite these disadvantages, Favarger based an observatory clock on this design which, when fitted with an invar pendulum and kept at a constant pressure, had an average daily rate of twenty-five milliseconds which was claimed to be comparable to the best astronomical clocks, such as those of Hipp and Riefler.[15]

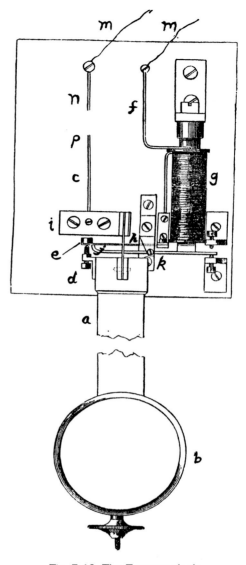

Fig. 7-10. The Froment clock.

Many attempts were made to improve Froment's design so that the impulse occurred at the mid point of the pendulum's travel but only two later examples will be mentioned here. The English Princeps clock[16], which also drastically reduced the time during which the current flowed in the coils, and that of Satori of Vienna[17] which improved the reliability of the electrical switching by resetting the spring through the contacts; a method pioneered by the Synchronome Company. The Satori clock also had the advantage that it required no lubrication and could therefore be used in an enclosure with the pressure reduced to between 5-10mm of mercury.

At the International Exhibition of Electricity in Vienna in 1884, August Joly displayed an electric clock which was impulsed through the suspension spring (ref.15 p. 265), possibly anticipating the use of a similar method by Riefler. This clock was not successful but it led to improved designs by Siemens of Berlin[18], Irk of Vienna[19] and Baumann of Furtwangen (ref. 15 p. 269). The arrangement of the Baumann clock is shown schematically in Fig. 7-11. The switching system, which is similar to that used by Hipp to obtain polarized seconds pulses from his observatory regulator (shown at I in Fig. 7-4B), reverses the current through the two solenoids as the pendulum passes through its mid point. This moves the soft-iron armature (which is polarized by the permanent magnets) in the opposite direction to that of the pendulum thereby giving the pendulum an impulse, in the correct sense, through the suspension strip.

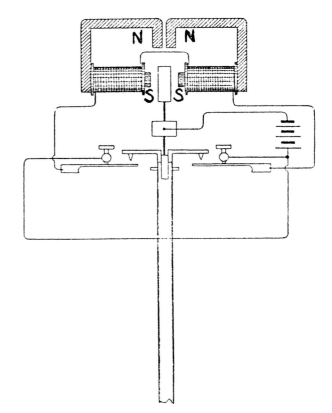

Fig. 7-11. The mechanism of the Baumann clock.

Gravity Remontoires

In 1852 an electric clock made by Charles Shepherd[13] was installed at the Royal Observatory at Greenwich (Fig. 7-12A). Its mode of operation can be seen from Fig. 7-12B. At the end of its swing to the left the pendulum moves the detent (ge) anticlockwise, releasing the gravity arm (dc) that falls on the screw (h) giving the pendulum an impulse to the right. At the end of its travel to the right the pendulum briefly closes a contact at (E), which energizes the two electromagnets, attracts the armature (a) and resets the gravity arm by means of the lever (bi). The dial on the clock and the two on either side of it are operated by currents of alternating polarity which are produced every second by the two contacts fitted at (l) and (k). The slave dials operate in the following way, as can be seen in Fig. 7-13 where the dial is mounted on the clock. Two magnets (Q) are mounted on an arbor (L) so that their poles are above the electromagnets (R, S). The electromagnets are wound in such a way that while (R) attracts one end of the permanent magnets, (S) repels the other. Thus when the current is reversed the magnets flip from the pole of one electromagnet to the other, in synchronism with the movement of the pendulum. The resulting oscillation of the arbor (L) is converted into a rotary motion by means of an escapement acting in reverse. Although the Shepherd clock provides a reasonably constant impulse it suffers from the same defects as the mechanical gravity escapement and in addition the pendulum has to make an electrical contact at the end of its travel. Nevertheless it was used as the Greenwich Mean Solar Standard clock from 1852 to 1893 and it provided the Greenwich Time Signal, which was relayed throughout the country. However it was corrected on a daily basis after comparison with the Sidereal Standard Clock. Shepherd was aware of the deficiencies of his clock and he proposed to impulse the pendulum at the center point of its swing[20] although no clocks with this modification appear to have survived. However the clock which he supplied in 1853, for use with the time ball at Deal, represents an improvement as the gravity arm is released electrically, reducing a source of friction. As can be seen in Fig. 7-14 the pendulum is mounted on the same plate as the dial and the same mechanism drives both. The oscillatory motion of the two magnets is transferred to the dial movement by means of a crutch and converted into a rotary motion as described above. The movement of the crutch also resets the gravity arm when the pendulum reaches the end of its swing in one direction and releases it at the end of its swing in the opposite direction.

Fig. 7-12A. The Shepherd mean solar clock at the Royal Observatory, Greenwich.

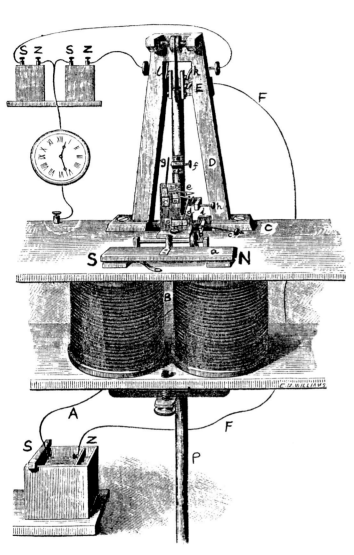

Fig. 7-12B. The mechanism of the Shepherd clock.

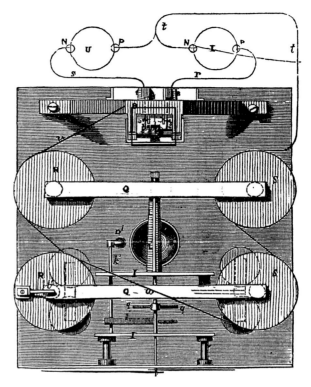

Fig. 7-13. The mechanism of the Shepherd slave dial.

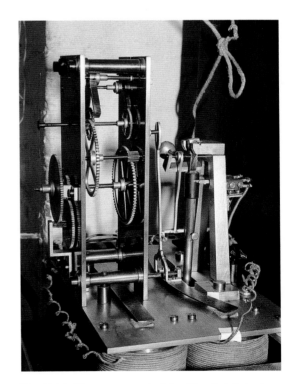

Fig. 7-14. The movement of the Shepherd clock associated with the Deal time ball.

151

In 1878 Shepherd patented an electric clock in which the oscillating magnets were replaced by magnets which rotated continuously in synchronism with the pendulum (Fig. 7-15).[21] This arrangement had the advantage that the inertia of the magnets would keep the clock going even if the impulsing current occasionally failed, although it had the disadvantage that it was not self-starting. Either this arrangement of rotating magnets was unsuccessful or Wheatstone[22] objected to its use, for three years later Shepherd patented an alternative design.[23] A clock made according to this patent was installed at Oxford University in 1883 and continued to work there for over seventy-five years (Fig. 7-16A). As can be seen from the close-up of the movement shown in Fig. 7-16B rotation is now obtained from the reciprocating motion of an S shaped permanent magnet. As in his other clocks, two contacts on either side of the pendulum produce pulses of current of alternating polarity that activate the two solenoids. The arms of the permanent magnet are therefore successively attracted and repelled by the solenoids and oscillate in synchronism with the pendulum. This motion is transferred to the crank on a vertical arbor (on the left-hand side of the figure) that rotates once for every two swings of the pendulum: in this instance once in every two seconds. The gravity arm is raised by a cam on this arbor and drops off it to impulse the pendulum. This is a desirable feature as it reduces the work which the pendulum must perform but there is evidence that the gravity arm was originally held by a detent and released by the pendulum, as shown in Fig. 7-15.

Fig. 7-16A. Shepherd clock, installed at the Examination Schools, Oxford University in 1883.

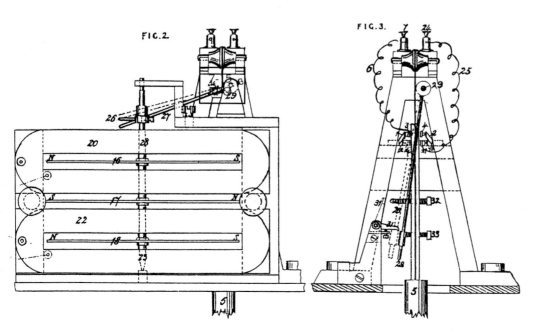

Fig. 7-15. Shepherd's clock with rotating magnets, from British Patent No. 2467, 1878.

Fig. 7-16B. The movement of the clock shown in Fig. 7-16A.

A rotary motion controlled by the pendulum is also used to release and reset the gravity arm in the clock which Alexander Steuart of Edinburgh patented in 1921.[24] In this instance the rotary motion is obtained from a continuously running direct current motor whose speed can be increased by short-circuiting a resistance; (r) in Fig. 7-17A. In this diagram the gravity arm (A) is impulsing the pendulum, which it continues to do until it arrives at the stop (R) and establishes electrical contact. This speeds up the motor rotating the cam (C) that lowers the arm (B) until it rests on its stop (T). In doing so it resets the gravity arm and breaks the electrical contact. If the motor is slow the duration of the contact will be extended to speed it up and vice versa. On its return swing to the left the pendulum picks up the gravity arm, which can now provide an impulse as the arm (B) is being moved away from it by the continuing motion of the cam. With a seconds pendulum the cam makes one revolution in two seconds and although its motion is not completely regular the variation is small. The motor can therefore be used to drive the motion work. The Steuart clock was made under license by the chronometer makers, Thomas Mercer (Fig. 7-17B). Given the slightly irregular motion of the motor it is surprising that the dial is calibrated in fifths of a second, but presumably they wanted to emphasis the advantages of continuous motion. A somewhat similar arrangement was used in the clock that the Earl of Meath and G. B. Bowell developed during the period 1933 to 1938.[25]

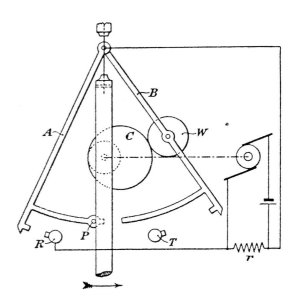

Fig. 7-17A. The mechanism of the Steuart clock

Fig. 7-17B. Steuart clock, as made by Thomas Mercer.

With the exception of the Meath-Bowell clock, all those mentioned in this section have relied on an electrical contact which is made at every swing or every other swing of the pendulum. The light contact necessary under these conditions can lead to unreliability and the freedom of the pendulum is violated by frequent contact with the gravity arm. During the early years of the twentieth century several makers used a count wheel[6] to extend the interval between the impulses delivered by the gravity arm and consequently that of the electrical contacts necessary to reset them. When combined with the so-called Synchronome switch[26] this produced a reliable master clock which dominated the British market for over fifty years. The example produced by the Synchronome Company is described here as it later formed an integral part of the Shortt-Synchronome free pendulum clock, although similar designs were produced by other makers, notably Gents. Fig. 7-18 shows the Synchronome movement in virtually its final form, as described by Hope-Jones in 1907.[27] As the pendulum moves to the right the click (B) advances the count wheel by one tooth. A vane (D) attached to this wheel releases the gravity arm (C) at half-minute intervals when the pendulum is approaching the mid-point of its swing. The impulse is given symmetrically about the mid-point of the pendulum's travel by the roller (R) running down the pallet (J). After it has delivered its impulse the tail of the gravity arm makes electrical contact with the armature (E) which is attracted by the electromagnet and begins to return the gravity arm to its original position. The movement of the armature is terminated by a stop but the momentum of the gravity arm carries it on until it is reset on the catch (K), also ensuring that there is a sharp break of contact. The virtue of this switch is that the energy to reset the gravity arm is transmitted through the contact surfaces and is derived from the power supply and not from the pendulum. When the patent had elapsed, it was used by Riefler, among others, to rewind their mechanical movements. Hope-Jones, the proprietor of the Synchronome Company, had attempted to shape the pallet (J) so that the impulse commenced gradually, and then built up to a maximum at the mid-point of the pendulum's travel before gradually decreasing. William Hamilton Shortt, in the discussion of a paper that Hope-Jones gave to the Institution of Electrical Engineers in 1910, derived the correct shape for the pallet mathematically.[28] However the performance of the clock still fell short of that of mechanical observatory clocks such as the Riefler; this was partly due to the variable friction in the movement. Shortt, who was employed as an engineer on the London & South Western Railway, decided to devote his spare time to improving the accuracy of the electric clock and his collaboration with Hope-Jones eventually resulted in the Shortt-Synchronome free pendulum clock.

Shortt began by improving the clock which Sir Henry Cunynghame and Hope-Jones had patented in 1907[29]. In this clock the impulses were delivered by a gravity arm that operated under the pendulum at every other swing. In 1911 Shortt produced an improved version with an inertia escapement,[30] which maintained a constant arc, but the results were not significantly better than those of the standard Synchronome (Fig. 7-18) and only about a dozen were produced. It transpired that half the energy supplied to the pendulum was being used to release the gravity arm and in an

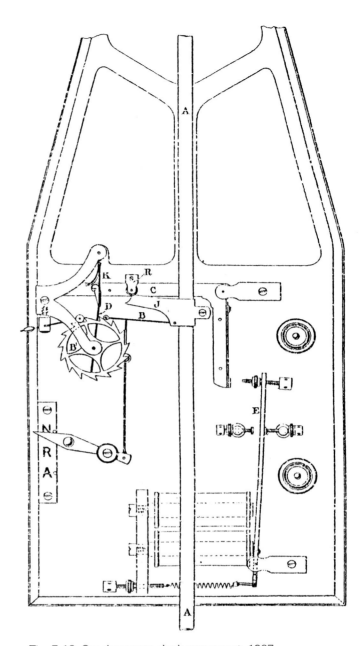

Fig. 7-18. Synchronome clock movement, 1907.

attempt to reduce the friction Shortt proposed to use two gravity arms, a light one to provide the impulse and a heavier one to make electrical contact. These improvements were incorporated in a clock[31] that Professor Sampson tested at the Royal Observatory, Edinburgh in 1915. After providing an impulse the lighter arm would release the heavier contact arm, which in falling would reset the impulse arm before being reset electrically through a Synchronome switch. This resulted in a delay of about three-quarters of a second between the release of the impulse arm and the establishment of the resetting contacts. Shortt reasoned that if the impulses were delivered each second the resetting contacts could also be used to release the impulse arms electromagnetically, provided they ran for a short time on dead faces. The pendulum would therefore swing freely apart from a brief period when it received an impulse. The mechanism of Shortt's Duplex clock, which incorporated this feature, is shown in Fig. 7-19; contact made by the left-hand arm (G.L) releases the right-hand impulse arm (B.R) by means of the electromagnet (S.R) and vice versa. The Duplex movement was found to have an excessive arc and attempts to reduce it by lightening the impulse arms led to uncertain unlocking of the contact arms. All Shortt's clocks had hitherto provided an electrical impulse directly every second, as he believed that this was essential for an observatory clock. He now accepted, what Hope-Jones had long advocated, that the impulses would have to be delivered less frequently, such as every half-minute, and that the seconds signals would have to be generated indirectly. There still remained the problem of increasing the delay from one to thirty seconds. Hope Jones believed that the best way of doing this was to use Rudd's concept of a free pendulum, maintained by impulses provided by a synchronized slave clock. A standard Synchronome clock could be used as a slave to impulse Shortt's clock if a means could be found for synchronizing it. This was achieved by a device known as the "hit and miss synchronizer," which Shortt patented in 1921,[32] although unknowingly he had been anticipated by W. S. Hubbard. Shortt used half the mechanism of his Duplex clock (shown in Fig. 7-19) to impulse the free pendulum at half-minute intervals by using the slave clock to release the impulse arm. The results were so promising that he installed the clock (with minor alterations) at the Royal Observatory in Edinburgh over Christmas 1921. Both pendulums were of course made of invar and the free pendulum was maintained at a low pressure in a cylindrical copper case. In February 1923 the Astronomer Royal for Scotland, Professor Sampson, informed Shortt that "the clock is unquestionably superior to Riefler, and Riefler never turned out a better clock than the one we have here." Shortt's clock, now known as SH.0, is preserved at the Royal Museums of Scotland in Edinburgh.

Shortt's design was modified slightly by the Synchronome Company with a view to producing it commercially and Hope-Jones presented the prototype, GC, to the Science Museum, London in 1935. The modification consisted of impulsing the pendulum near the top instead of underneath the bob, which greatly simplified the construction and reduced further the friction in the movement. Fig. 7-20A shows the layout of the clock, in the form in which it was produced by the Synchronome Company. On the right-hand side is a conventional Synchronome clock, which in this instance serves as the slave. In addition to advancing the slave dial the half-minute electrical impulse also energizes the electromagnet (c) which releases the gravity arm (d) of the master or free pendulum as it approaches the central position moving towards the left. A semicircular jewelled pallet with the flat side facing downwards is mounted on the end of the impulse or gravity arm and falls on the dead face of the light impulsing wheel that is attached to the pendulum. As the pendulum moves further towards the left, the jewelled pallet travels down the wheel imparting a gradually increasing impulse about the centre-point of the pendulum's travel. Rather surprisingly Shortt made no attempt to produce a symmetrical impulse, as he did in deriving the shape of the pallet of the standard Synchronome clock, although as Phillip Woodward has pointed out this lopsidedness is of no consequence.[33] When the gravity arm eventually drops off the wheel, its tail releases the contact arm (l), which in falling resets the gravity arm. The contact arm is reset through the Synchronome switch, which also advances the master dial and activates the "hit and miss synchronizer" (g). This provides information about the position of the free pendulum so that the pendulum of the slave clock can be synchronised with it. There is a delay of about 0.8 seconds between the release of the free-pendulum gravity arm by the slave clock and the operation of the hit and miss synchroniser. During this time the slave pendulum will have completed its swing to the right and arrived at the mid point of its swing to the left. If the two pendulums are in the correct phase to each other the armature of the hit and miss synchroniser (which has been attracted by the electromagnet) will not interact with the vertical spring (k) mounted on the slave pendulum. However this pendulum is adjusted so that it looses about six seconds a day (or 1/480 sec per half minute) relative to the free pendulum so that it will arrive at the hit and miss synchroniser after the armature has descended. As the pendulum continues to swing to the left the spring (K) is bent, increasing the restoring force on the pendulum and causing it to gain about twice the amount which it lost during the previous half-minute (1/240 second). Thus the hit and miss synchroniser will interact with the pendulum approximately every other time it is activated. This maintains the phase relationship between the two pendulums to within +/- 1/240 second so that the slave clock is able to release the gravity arm (d) at the right time to impulse the free pendulum. The pendulum swings freely for 99% of the time; all the other functions are carried out by the slave clock. The variation in the daily rate was +/- 2 millisecond.

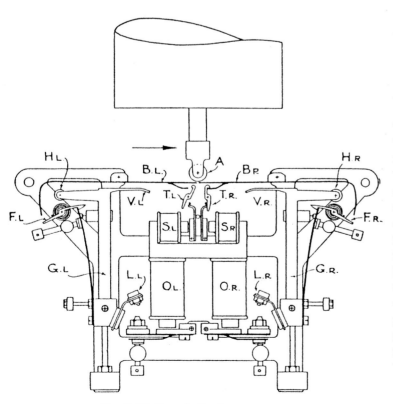

Fig. 7-19. Shortt's Duplex movement.

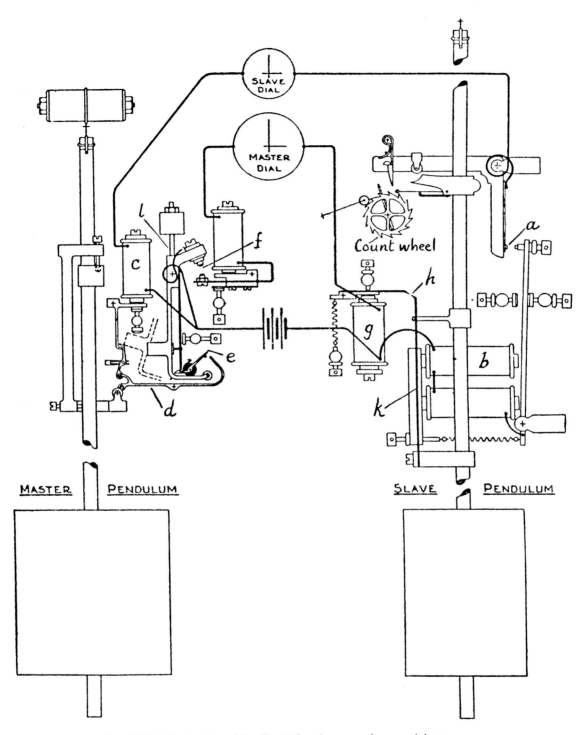

Fig. 7-20A. Mechanism of the Shortt Synchronome free pendulum clock.

The first Shortt-Synchronome free pendulum clock was sold to Helwan Observatory in Egypt in 1924 and over the next thirty-odd years around a hundred were sold to observatories and laboratories throughout the world. The clocks remained basically unchanged during this long production run, as can be seen from the example shown in Fig. 7-20B, which was sold to the Soviet Union in 1951. The free pendulum on the left is housed in a copper cylinder to maintain it at a constant low pressure. In practice it had been found that the period of the Shortt pendulum is most stable when the pressure is between 15 to 20mm of mercury.[34] The slave clock on the right has two small dials below the main dial, which correspond to the half-minute slave and master dials show in Fig. 7-20A and there will consequently be a delay of about 0.8 seconds between their operation. The large regulator dial is driven by seconds impulses which are produced by a subsidiary Synchronome switch which is released at every swing of the slave pendulum.

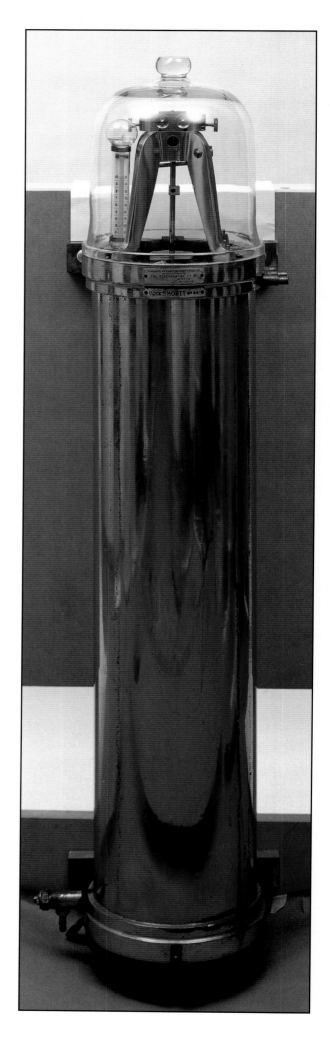

Fig. 7-20B. Shortt-Synchronome free-pendulum clock No. 84, supplied to the Soviet Union in 1951, together with its slave clock shown in Fig. 7-20C.

Fig. 7-20C.

The Synchronome Company supplied more free-pendulum clocks to the Soviet Union than to any other country and it is hardly surprising that they decided to produce their own version. I. I. Kvarnberg started to work on this clock at the All Union Scientific Research Institute of Metrology in 1934 and two years later the Etalon plant in Leningrad was testing the first models. This work was interrupted by the Second World War and only completed in 1952 by which time it was claimed that the variation in daily rate had been reduced to one millisecond,[35] half that of the Shortt. Production probably started shortly afterwards as the last Shortt-Synchronome clocks were supplied to Russia in 1951. An Etalon free pendulum clock made in 1955 is shown in Fig. 7-21.

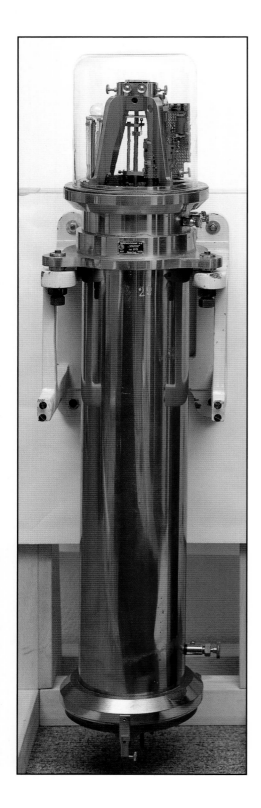

Fig. 7-21. Etalon free-pendulum clock No. 27, made in Leningrad in 1955, together with its slave clock.

Postscript

With one or two notable exceptions the electric clock was at first viewed with suspicion and it was not generally accepted as a reliable timepiece until the early years of the twentieth century. A further twenty years was to elapse before it was able to challenge the supremacy of the mechanical clock in the observatories of the world. The Royal Observatory at Greenwich purchased three Shortt-Synchronome free-pendulum clocks between 1924 and 1927, where they became the sidereal and mean-time standards. One clock ran for almost nine years without interruption, demonstrating its reliability, and the error of another was never more than half a second over two and a half years, demonstrating its long term stability.[36] However their supremacy was short-lived for in 1927 Marrison patented the quartz crystal clock.[37] Although its performance in the short term was superior to that of the Shortt clock, a drift in its rate due to the crystal ageing made the Shortt clock the preferred timekeeper for periods longer than one day. However this only offered a temporary respite as the ageing of the quartz crystal was gradually overcome and by 1942 quartz clocks had replaced the Shortt clocks as the standard timekeepers at the Royal Observatory. Despite losing their position as the ultimate timekeepers there was still a demand for robust and easily managed precision pendulum electric clocks, such as the Shortt, Etalon and Fedchenko. Production of the Shortt continued until 1956 and production of the Fedchenko ceased in 1970.

References

[1] British Patent No. 10,838, "Improvements in Electric Clocks and Telegraphs," 1845.

[2] Bain, Alexander, *A Short History of the Electric Clocks...* (London: Chapman, 1852; repr. London: Turner & Devereux, 1973).

[3] British Patent No. 9745, "Certain Improvements in Producing and Regulating Electric Currents", 1843.

[4] Shenton, F. G. A., "The Earth Driven Clock," *Antiquarian Horology*, 8 (December 1972), 63.

[5] *Electrical Review*, 69 (1911), 619.

[6] British Patent No. 4882, "Electric Time-pieces," 1881.

[7] Hirsch, A., *La Pendule électrique de précision de Hipp* (Neuchâtel: 1891). For an English translation see Aked, C. K., "Electrical Precision", *Antiquarian Horology*, 14 (June 1983), 172.

[8] Cornu, A., "Sur la synchronisation des horloges de précision et la distribution de L'heure," *Comptes rendus de l'Académie des Sciences*, CV.2 (1887) 1106. Etc.

[9] Lavet, M., "Les progrès récents de l'Hologerie électrique," *Revue générale de l'électricité* (1922), 845.

[10] Féry, C., "Les Piles Employées en Horlogerie Électrique," *Annales Françaises de Chronométrie*, 4 (1934), 165.

[11] British Patent No. 746,465, "Improved Electromagnetic Impulse Device for Driving Clocks," 1954.

[12] "A New High-Precision Clock", *Horological Journal*, 77 (1935) 312, 332, 359; 78 (November 1935) 8; 78 (December 1935) 3.

[13] British Patent No. 12567, "Improvements in working clocks and other timekeepers etc by electricity," 1849.

[14] Du Moncel, Th., *Exposé des Application de l'Électricité*, I (Paris: Hachette, 1855), p. 99. Figuier, L., "Les Horloges Electriques," *Le Mervielles de la Science*, II (Paris: Furne, Jouvet, ?1868), 413.

[15] Favarger, A., *L'Électricité et ses Applications a la Chronométrie*, 3rd edn (Neuchâtel: 1924), p. 264.

[16] British Patent No. 206,186, "Improvements in Electric Clocks," 1922. Prince, C. E., "Some Recent Developments in Electric Clocks," *Horological Journal*, 67 (1922) 57, 83.

[17] Novak, E. "Pendelantrieb von Satori," *Deutsche Uhrmacher Zeitung*, (1933).

[18] Deutsches Reich Patent No. 204,351, 1907.

[19] Krumm, G., *Die elektrischen Uhr*, (Bautzen: Hübners Verlag,), p. 263.

[20] Denison, E. B., "Clock and Watch Work" in *Encyclopaedia Britannica*, 8th ed. (Edinburgh: Blackie, 1854), VII, p. 27.

[21] British Patent No. 2467, "Improvements in Electro-magnetic Clocks," 1878.

[22] Wheatstone had used a similar arrangement to drive his slave dials, British Patent No. 3028, "Electro-magnetic Clocks etc.," 1869.

[23] British Patent No. 369, "Electro-magnetic Clocks and Batteries for the same etc.," 1881.

[24] Steuart, A., "An Electric Clock with Detached Pendulum and Continuous Motion," *Proc. Roy. Soc. of Edinburgh* (1922-23), 154.

[25] G. B. B[owell], "The Zero Escapement of the Meath Clock," *Horological Journal*, 80 (1938), 12.

[26] British Patent No. 1587, "Improvements in or appertaining to Electric Clocks and similar instruments," 1895

[27] Hope-Jones, F., "Cunynghame's Detached Gravity Escapement," *Horological Journal*, 50 (Dec 1907), 65.

[28] Hope-Jones, F., "Modern Electric Time Service," *Journal of the Institution of Electrical Engineers*, 45 (1910), 97.

[29] British Patent No. 1945, "Improvements in Gravity Escapements and Electric Clocks", 1907

[30] Shortt, W. H., "Precision Timekeeping", *Horological Journal*, 54 (1912), 117.

[31] Shortt, W. H., "Some Experimental Mechanisms, Mechanical and Otherwise, for the Maintenance of Vibration of a Pendulum", *Horological Journal*, 71 (1929) 225.

[32] British Patent No. 187814, "Improvements in the Synchronisation of Clocks," 1921.

[33] Woodward, P., *My Own Right Time* (Oxford University Press, 1995), p. 34.

[34] Loomis, A. L., "The Precise Measurement of Time", *Monthly Notices of the Royal Astronomical Society*, 41 (1931), 573.

[35] Ermakov, V. I., Pushkin, S. B. & Sachkov, V. I., "State Standards of Time Measurement and Frequency", *Measurement Techniques* (1967), 1352.

[36] Dyson, F., in the *Watch and Clockmaker* (1931) 238.

[37] US Patent No. 1,788,533, 1927.

General Works

Hope-Jones, F., *Electric Clocks* (London: N.A.G. Press, 1931).

Hope-Jones, F., *Electrical Timekeeping* 2nd edn (London: N.A.G. Press, 1949; repr. 1976).

Favarger, A., *L'Électricité et ses Applications a la Chronométrie*, 3rd ed. (Neuchâtel: 1924).

Guye, R. P., & Bossart, M., *Horlogerie Électrique*, (Lausanne: Scriptar, 1948).

Chapter 8
Pendulum Clock Precision 1750-1960
BY A. D. STEWART

Most of this chapter concerns clocks in astronomical observatories, the annual publications of which frequently contain data on the rates and accumulated errors of the transit clock. These rates and errors can be used, as described below, to estimate the clock's precision. The main thrust of the chapter, however, is the identification of the reasons for the marked improvement in clock precision from the eighteenth century onwards.

Precision is an intrinsic characteristic of a clock, quite unrelated to accuracy. If every beat of a pendulum takes exactly the same time then it is, by definition, absolutely precise. The beat of real pendulums, especially those connected to clockwork, is never regular, a fact most easily expressed by simply calculating the standard deviation of clock rates. The classical method of rating a clock is to observe the time a named star passes behind a fixed point each sidereal day. The measured differences in the length of each successive day give the daily rate. The standard deviation describes the dispersion of the rates, all slightly in error, about the mean value. It defines the precision of the clock, but gives no help in getting it to tell the time of day.

The accuracy of clock time is measured by its degree of conformity with an extrinsic, independent standard, now usually coordinated universal time (UTC). The U.S National Institute of Standards and Technology web site (http://www.boulder.nist.gov/timefreq/general/glossary.htm) defines precision, accuracy, UTC and much else connected with time and frequency.

An early example of a precision calculation, originally published by the Riefler Company, is shown in Fig. 8-1. The clock Riefler #169, built in 1906, had an invar pendulum that hardly responded to temperature changes, with a barometric coefficient of -0.009 seconds per day per millibar change in pressure (s/d/mbar). It was housed in a dust-proof but not air tight case. Column 7 of the Table gives the observed daily rate of the clock (the plus sign indicates losing), while column 8 gives the rate after correction for the changing barometric pressure. Column 9 gives the daily rate less the average rate and column 10 the same values squared. The sample standard deviation is then the square root of the sum of the squares in column 10 divided by $(n - 1)$, where n is the number of daily observations. Columns 9 and 10 are nowadays redundant for one can instantly obtain the sample standard deviation (= 0.021 s) by simply tapping the data of column 8 into a pocket calculator.

1	2	3	4	5	6	7	8	9	10
1906 Dec.	**Pressure** at 10 a.m.	**Pressure** average	b = av. press. −716 mm	b × 0.012	Accum. error	Observed daily rate	Reduced daily rate	Average daily rate − daily rate Δ	Δ²
	mm	mm	mm	mm	seconds	seconds	seconds	seconds	seconds
21	727				+1.61				
		726	+10	−0.12		+0.18	+0.06	−0.01	0.0001
22	725				+1.79				
		724	+8	−0.10		+0.17	+0.07	−0.02	0.0004
23	723				+1.96				
		723	+7	−0.08		+0.11	+0.03	+0.02	0.0004
24	723				+2.07				
		717	+1	−0.01		+0.05	+0.04	+0.01	0.0001
25	711				+2.12				
		708	−8	+0.10		−0.08	+0.02	+0.03	0.0009
26	705				+2.04				
		703	−13	+0.16		−0.12	+0.04	+0.01	0.0001
27	701				+1.92				
		703	−13	+0.16		−0.08	+0.08	−0.03	0.0009
28	705				+1.84				
		706	−10	+0.12		−0.06	+0.06	−0.01	0.0001
29	707				+1.78				
					Total =		+0.40 s	0.14 s	0.0030
					Average daily rate =		+0.05 s		
					Average variation in daily rate =			±0.017 s	
					Standard deviation of daily rate =	$\sqrt{[(0.0030)/(8-1)]}$			= ±0.021 s

Fig. 8-1. **Table 1: Eight daily rates of the Riefler clock #169.** The table shows how to correct the rates for changes in barometric pressure, and calculate their standard deviation.

The standard deviation, however, is unsatisfactory for long series of daily rates. Imagine a clock that gains steadily for fifteen days at 0.30 s/d and then loses steadily at the same rate for another fifteen days. The standard deviation of the daily rates measured over the first *ten* days would be zero, since all the rates are the same, but over the second interval of *ten* days it would be 0.316 s despite the fact that the clock over periods of day continued to be extremely precise.

A solution to this problem was provided by Allan *et al* in 1974.[1] A simple description of the procedure is given by Woodward.[2] Allan *et al* proposed that instead of summing the squares of the differences from the mean, one should sum the squares of the differences *between successive rates*. If this sum is s and the number of differences is n then a characteristic error ε can be calculated:

$$\varepsilon = \sqrt{[s/2n]} \text{ seconds}$$

For example, the error ε for the set of observations shown in Table 1 comes out at 0.018 s. For short series of data, in fact, the error ε and the sample standard deviation turn out to be numerically very similar.

The daily rates shown in Table 1 were obtained by comparison with a regulator clock in a constant pressure container, but in old observatories the usual practice was to establish the passage of time by reference to star transits. In the eighteenth century this was done at weekly intervals, weather permitting, and the rate assumed to be absolutely stable in between. The assumed rate was then used to correct times noted during the week. Data of this kind allow the estimation of ε based on the error that accumulated every eight days, conveniently called $\varepsilon(8)$ to distinguish it from $\varepsilon(1)$ based on daily rates. In some cases only $\varepsilon(16)$ can be estimated. Nineteenth century observatory clocks were checked more frequently and values for $\varepsilon(1)$ can usually be obtained. In the absence of suitable data $\varepsilon(1)$ cannot, unfortunately, be found by simply dividing $\varepsilon(8)$ by eight, for reasons that have been explained by Woodward.[2]

The selection of observatory clock data presented in Table 2 (Fig. 8-2) is intended to illustrate the evolution of precision over the centuries. It is based on sets of rates collected while the clocks were in use. Note, however, that the precision of any clock varies significantly from year to year, and that clocks nominally of the same design but operating in different conditions may have quite different precision. The following notes should be read in conjunction with Table 2.

One of the first regulator clocks for which records are available is that built by George Graham for the Greenwich Royal Observatory in 1750.[3] The clock had a dead-beat escapement and a pendulum with Harrison's gridiron compensation.[4] Although temperature and barometric pressure were recorded at Greenwich at this time they were not used to correct the clock rates. A list of rate changes that occurred over thirty-four different days during 1762, 1763 and 1764 has an average of 0.4 s, giving a rough indication of $\varepsilon(1)$.

Arnold's regulator at Mannheim Observatory also had a dead-beat escapement, and a five-bar zinc and steel pendulum. The clock is now in Heidelberg Observatory and has been described and figured by Mercer.[5] Multiple regression analysis of data for this clock, originally published by Mayer in 1781 and recently reprinted by Staeger[6], gives a barometric coefficient of -0.025 s/d/mbar and a temperature coefficient of -0.065 s/d/°C. Table 2 shows values for ε both before and after correction for pressure and temperature variations.

Clock	Built	Year observed	Starting date	Days analysed	$\varepsilon(1)$ s	$\varepsilon(8)$ s	$\varepsilon(16)$ s	$\varepsilon(32)$ s
Graham (Greenwich)	1750	1756	8–1	378	–	0.9	1.6	2.5
Arnold (Mannheim)	1779	1779	9–1	132	–	3.2	–	
	1779	1779	9–1	132	–	–	**2.6**	–
Hardy (Greenwich)	1809	1820	4–22	206	–	0.9	1.6	4
Hardy (Cambridge)	1825	1833	4–28; 9–16	60	**0.05**	0.5	–	–
Knoblich #1770 (Bothkamp)	1868	1911	2–1	189	–	0.17	–	–
Dent #1906 (Greenwich)	1870	1910	1–24	38	0.06	–	–	–
		1910	1–17; 9–24	216	–	0.4	–	–
		1910	1–10	350	–	–	0.9	5
Hipp (Neuchâtel)	1884	1889	12–26	371	–	0.2	–	1.2
Morrison #8702 (Teddington)	1901	1906	11–13	205	–	0.2	0.6	–
		1935	3–1	90	**0.03**	–	–	–
Riefler #254 (Paris)	1911	1917	10–20	388	–	0.1	–	0.2
Shortt #13 (Teddington)	1925	1958	11–11	86	**0.006**	0.05	–	–
Riefler E-type (Nesselwang)	1960	1960	5–4	52	**0.002**	–	–	–

The starting date of each series of observations is given as month–day. Precision shown in bold type has been corrected for pressure variations, or has been obtained from clocks in sealed containers.

Fig. 8-2. **Table 2: The precision of some observatory pendulum clocks 1750-1960.**

In view of the controversy surrounding the quality of Hardy's regulators the ϵ values for two of them are given in Table 2. Both had temperature compensated pendulums and his spring pallet escapement. The Greenwich data published by Pearson in 1829[7] give values for $\epsilon(8)$, $\epsilon(16)$ and $\epsilon(32)$ almost identical to those for the Graham dead-beat regulator in the same observatory (see above). The adopted daily rates for the Cambridge clock[8] for the periods examined show no correlation with barometric pressure even though it several times changed by 20 mbar and once by as much as 30 mbar. The daily rates had evidently been corrected for pressure variations, though this is not actually stated in the cited publication. A Hardy regulator said to be like that installed at Greenwich and recently subjected to a 40 day trial gave $\epsilon(1) = 0.08$ s after correction for barometric and temperature errors.[9] This is quite similar to the value obtained for the Cambridge clock.

The clock Knoblich #1770 at Bothkamp Observatory had a pendulum almost perfectly compensated for temperature and pressure variations, and operated at nearly constant temperature. It probably had a dead-beat escapement. Further details of this clock will be found later in the chapter.

The Dent clock at Greenwich observatory was described and figured in the journal *Nature* soon after it was installed.[10] The pendulum had mercurial compensation and a magnetic device for correcting the effect of variable atmospheric pressure. The escapement was unusual. It had been proposed by Airy in 1827 and may be described as a detached spring-detent escapement, with a single pallet that gave impulse to the one-second pendulum every two seconds. Despite the originality of the escapement the estimates of $\epsilon(1)$, *etc.* based on data recorded in 1910[11], are by no means remarkable.

The Swiss clockmaker Matthaus Hipp (1813-1893) devised a method of electro-mechanically impulsing clock pendulums in 1834, and by 1881 had developed it for use in observatory regulators.[12] The clock supplied to Neuchâtel Observatory in 1884 operated in an air-tight glass bell at a reduced pressure of about 60 mbar. The mercury compensated pendulum had a barometric coefficient of -0.009 s/d/mbar, and a temperature coefficient of -0.005 s/d/° C. The change in pressure within the glass bell over a year was only 5 mbar, with minimal effect on the rate of the clock. How the pressure came to be so constant, despite annual temperature changes of over 15° C, is a mystery, for there is no record of a pump being used. The pendulum was impulsed very near the centre point of its oscillation whenever the arc dropped to a predetermined value. This occurred about once a minute. The rates were determined once a week and so $\epsilon(7)$ is shown in place of $\epsilon(8)$ in Table 2 (Fig. 8-2).

The Morrison clock at Teddington was built to serve as the standard for the newly established National Physical Laboratory in 1901. Its construction and performance have been described by the author.[13] The clock is a dead-beat regulator of standard design, but was originally housed in an air-tight case with the pressure kept at about 960 mbar. The original pendulum was a Riefler J-type with a temperature co-efficient of under 0.005 s/d/° C. The record of this clock is good, but would have been even better had it been possible to keep it at a more constant temperature (and hence pressure). In this case it is probable that $\epsilon(1)$ would have been about 0.02 s, rather than the 0.03 s shown in Table 2.

The Riefler clock #254 was also kept in an air-tight case, at a reduced pressure of about 800 mbar. The escapement was the well-known spring remontoire design. The invar pendulum was designed to cope with temperature stratification in the case as well as overall temperature changes. The record of this clock over the period 1917-1918, originally published by Kienle, has been reprinted by Riefler.[14] The interval between observations is close to ten days, so that the value $\epsilon(10)$ is shown in place of $\epsilon(8)$ in Table 2. There are no data which would allow an estimate of $\epsilon(1)$ but the standard deviation of daily rates for clocks of this design was warranted by the maker to lie in the range 0.01-0.03 s.

The Shortt clock #13 had a "free" master pendulum made of invar, impulsed electro-mechanically every 30 seconds, swinging in an air-tight case at a pressure of 33 mbar. The clock was in use at the National Physical Laboratory in Teddington when the observations were made, to an accuracy of 0.001 s/d, between November 1958 and February 1959.[15] The record of another Shortt clock, #41 at the U.S. Naval Observatory in Washington, D.C. is, however, significantly better, perhaps because it was mounted on a more stable foundation. The air pressure in the case was 20 mbar when the clock was restarted in 1984 after many years of inactivity and timed to an accuracy of 0.00001 s/d for a year.[17] The values of $\epsilon(1)$ and $\epsilon(8)$ from these measurements are 0.0008 s and 0.0124 s respectively.[16,18]

The Riefler E-type was housed in an air-tight case with the pressure maintained at 66 mbar. The movement was fitted with Riefler's gravity escapement, patented in 1913, and a super-invar pendulum. The pendulum arc was stabilised to ± 0.001°. The clock was claimed to have a standard deviation in daily rate normally of 0.004 s and over short periods of only 0.0001 s. The daily rates analysed here were determined to an accuracy of ± 0.0005 s in the Riefler Company's clock vault at Nesselwang.[19] It is not possible to get a reliable estimate of $\epsilon(8)$ from this set of data for it was purposely selected to show the clock's slow response of rate to a slight earthquake shock.

The data for Riefler E-type and Shortt clocks are comparable with those for the Fedchenko clock, manufactured in the former Soviet Union. The most advanced version, called AchF-3, first produced in 1958, had a single invar pendulum which was electro-magnetically impulsed every two seconds, with a suspension spring designed to ensure isochronality despite arc fluctuations. The pressure in the air tight case was only 5 mbar. The standard deviation of daily rate of AchF-3 was said to be 0.0002-0.0003 s.[20]

The production of mechanical clocks for the generation of standard time ceased about 1960 in the face of competition from significantly cheaper quartz and atomic clocks.

The progressive improvement in precision through the centuries shown in Table 2 was partly due to a better appreciation of the need for a thorough analysis of clock rates, thermal stability and a vibration-free environment. This can be illustrated by reference to two observatory regulators - Graham's at Greenwich in 1756 and the Knoblich clock at Bothkamp Observatory, near Kiel, in 1914.

The Graham regulator (see Table 2) stood on a wooden floor in the transit room, exposed to large, abrupt temperature changes whenever the roof shutters were opened for stellar observations. No correction was made for barometric pressure changes, nor for any lack of temperature compensation by the gridiron pendulum. Times observed were simply corrected retrospectively every week or so. If, say, the clock gained a second between transit observations ten days apart it was assumed to have gained exactly 0.1 s/d, and intervening observations adjusted accordingly.

Compare this with the care lavished on the clock Knoblich #1770, built about 1868, the performance of which was described by Schiller in 1914.[21.] The clock probably had a dead-beat escapement, though this is not actually stated in the paper. It had a mercury compensated pendulum with a mercury manometer to compensate for varying atmospheric pressure. The case was insulated so that daily temperature changes within it were less than 0.3°C. Three thermometers and a hair hygrometer were mounted within the case. The thermometers measured the temperature stratification, while the hygrometer served to measure the relative humidity, kept at 20%. Although the clock rates were satisfactorily stable it was decided to use multiple regression analysis to determine the coefficients for average temperature, barometric pressure and temperature stratification, using data collected over a period of twenty-six months - a laborious task considering that at the time there were neither computers nor pocket calculators. During most of the 26 months the rates were determined at irregular intervals but during the last 189 days, from 1 February 1911, they were measured every eight days, allowing $\varepsilon(8)$ to be estimated at 0.17 s. The redetermined regression coefficients for this period of 189 days turn out to be very small, so that $\varepsilon(8)$ from the corrected rates (0.16 s) is hardly any less than from the uncorrected ones. In other words the compensation of the pendulum was almost perfect and the temperature stratification had little evident effect. There are no data which would allow calculation of $\varepsilon(1)$. However, a twenty-five day record for Knoblich #1952 in the astrophysical observatory at Potsdam, built about 1877 and observed in 1878, gives $\varepsilon(1) = 0.05$ s. This clock had a compensated pendulum like that at Bothkamp, and a dead-beat escapement.[22]

It is worth considering at this point what performance could be expected from the old Graham clock at Greenwich if it were treated with the same care as the Bothkamp regulator. The first correction that should be made to the observed rates is for fluctuations in barometric pressure, for which the barometric coefficient is needed. Fortuitously, the author has in his possession a regulator with a pendulum that appears to be like that of the Graham clock. It has a Harrison gridiron and a lenticular bob, with a total mass of 11.3 kg. The barometric coefficient is -0.03 s/d/mbar. The magnitude of atmospheric pressure fluctuations in the south of England documented by Woodward[23] combined with this coefficient predict, for an otherwise error-free clock, $\varepsilon(1) = 0.16$ s and $\varepsilon(8) = 2$ s. In other words, nearly half the daily error and all the weekly error of the Graham clock shown in Table 2 can be attributed to the ups and downs of atmospheric pressure. If, in addition, the daily temperature variation could be reduced a further increase in precision can be anticipated, for gridiron pendulums are not always well compensated. By these comparatively simple means George Graham's clock might be made a match for those by later makers, even Dent #1906 at Greenwich.

If clocks with dead-beat escapements can produce such good results one might wonder at Riefler's undoubted success with the spring remontoire escapement. The answer appears to lie not so much in the superiority of the Riefler escapement but rather that the Company guaranteed the performance of their clocks from the moment of installation, whereas the purchaser of a clock by any other maker could look forward to years of effort to bring it to an acceptable level of precision. A good example is given by the Hipp clock at Neuchâtel, described earlier. The Director of the Observatory wrote in 1891: *It would be of little use to go into the details of the performance during the years 1884 to 1888, which included a number of experiments and the correction of the regulation; it is not until the end of July 1888, when the last regulation of the compensation had been made, that it is of interest to study the performance of the clock.*[12]

By 1888, in fact, the precision of the Hipp clock at Neuchâtel was three times what it had been when it was installed four years earlier.

As we have seen, the care bestowed on a clock is critical in achieving high precision, but there is also another important factor - the energy loss from the pendulum during each stroke. Riefler and other makers had started encasing their clocks in air-tight chambers around 1890, initially with the idea of stabilizing the air pressure. But it later became evident that reducing the pressure in the chamber produced an unexpected improvement in rate stability. Although there was no adequate theoretical explanation for this at the time, the facts are clear. A Riefler clock with a first class (type A) movement and a J-type invar pendulum working in an airtight chamber at reduced pressure of 750-800 mbar was claimed by its maker to have a standard deviation of daily rate in the range 0.01-0.03 s, as compared with 0.03-0.06 s for the same clock at atmospheric pressure.[24] Thus, by simply using a hand pump the precision had been increased between two or three times. In the 1960s, at the end of the mechanical clock era, pressures were only a few tens of millibars and the precision was ten to a hundred times better than that achievable at atmospheric pressure. It is impossible to show with presently available data that any of this improvement in precision derived from the mode of impulsing the pendulum (*e.g.* an escapement), even though there are good theoretical reasons for expecting it. Rather, the key factor appears to be the proportion of the pendulum's total energy lost during each stroke.[25] If this is very small, as when the pendulum swings in a near vacuum, the only energy absorbed is by the suspension, and the pendulum will swing virtually for ever without impulse. The precision in such circumstances is high.

References

[1] Allan, D.W., Shoaf, J.H. & Halford, D., 1974. *Statistics of time and frequency data analysis*. National Bureau of Standards Monograph 140, pp. 151-187.

[2] Woodward, P. 1993. *Measurement of clock performance, in A. L. Rawlings, The science of clocks and watches*. British Horological Institute, Upton Hall, pp. 353-368.

[3] Bradley, J. 1805. *Astronomical observations made at the Royal Observatory at Greenwich from the years MDCCL to the year MDCCLXII*. vol. 2 Oxford, pp. 308-309.

[4] Howse, D. 1975. *Greenwich Observatory, 3: The buildings and instruments*. Taylor & Francis, London, p. 129.

[5] Mercer, V. 1972. *John Arnold & Son chronometer makers 1762-1843*. The Antiquarian Horological Society, London, pp.112-116.

[6] Staeger, H. 1997. *100 years of precision timekeepers from John Arnold to Arnold & Frodsham*. Filderstadt, pp. 746-748.

[7] Pearson, W. 1829. 'Practical astronomy', vol. 2, 313.

[8] Airy, G.B. 1834. *Astronomical observations made at the Observatory of Cambridge. Vol. VI for the year 1833*. Cambridge. Table of observed transits.

[9] Redfern, J. & Woodward, P., 1993. *The performance of a 19th century regulator by William Hardy*. Horological Journal 135, pp. 306-312.

[10] Anon, 1875. *The new standard sidereal clock of the Royal Observatory, Greenwich*. Nature 11, pp. 431-433.

[11] Christie, W.H.M. & Dyson, F.W. (eds) 1912. *Results of the astronomical observations made at the Royal Observatory, Greenwich, in the year 1910*. Edinburgh, Table IV.

[12] Aked, C.K., 1983. *Electrical precision*. Antiquarian Horology 14, pp. 172-181.

[13] Stewart, A.D. 1997. *The Morrison clock at the National Physics Laboratory*. Horological Journal 139, pp. 242-244.

[14] Riefler D. 1991. *Riefler-Präzisionspendeluhren 1890-1965*. Callwey, München, 2nd edition, p. 161.

[15] Burch, F.H. 1960. *How accurate is a clock?* Horological Journal 102, p. 232.

[16] Boucheron, P.H. 1986. *Effects of the gravitational attraction of the sun and the moon on the period of a pendulum*. Antiquarian Horology 16, pp. 53-65.

[17] Boucheron, P.H. 1987. *Tides of the planet Earth affect pendulum clocks*. Bulletin of the National Association of Watch & Clock Collectors 29, pp. 429-433.

[18] Woodward, P. 1995. *My own right time: an exploration of clockwork design*. Oxford University Press, Oxford, pp. 124.

[19] Riefler, *op. cit.*, 102, Fig. 137.

[20] Feinstein, G. 1995. *F.M. Fedchenko and his pendulum astronomical clocks*. Bulletin of the National Association of Watch & Clock Collectors 37, pp. 169-184.

[21] Schiller, K. 1914. *Untersuchung über den Gang der Hauptuhr der Bothkamper Sternwarte Knoblich 1770*. Astronomische Nachrichten 198, pp. 90-94. Reprinted in Riefler, *op. cit.*, 155-161.

[22] Becker, E. 1880. *Ueber den Gang der Penduluhr Knoblich Nr. 1952*. Astronomische Nachrichten 96, no. 2290, pp. 151-156.

[23] Woodward, P. 1987. *Stability analysis, a branch of modern horology, part 3*. Horological Journal 129, pp. 15-16.

[24] Riefler, *op. cit.*, pp. 84-97.

[25] Bateman, D.A. 1977. *Vibration theory and clocks*. Part 3. Horological Journal 120, pp. 48-55. *Vibration theory and clocks*. Part 7. Horological Journal 120, pp. 56-64.

Chapter 9
Thomas Tompion and George Graham
by Derek Roberts

Although Tompion and Graham worked together for some seventeen years, the last three of which were in partnership, little is known of their relationship, other than that it was a cordial one.

Undoubtedly in his early days Graham must have learnt much from Tompion, the leading horologist and scientific instrument maker of the day and it is also likely, particularly at a later period, that he became acquainted with many of the eminent scientists who visited Tompion and gained further knowledge from them also. Unfortunately however, no work books or other correspondence relating to the business exist and thus we shall never know what ideas passed between them relating to escapements, repeating work and a host of other matters and can thus only speculate on such things as the dead beat escapement.

It is interesting that their lives trod somewhat similar paths. Both had a keen interest in scientific instruments but relied on a standard series of clocks and in particular watches for their bread and butter, Graham especially making use of the same designs over a remarkably long period of time. Undoubtedly both were considered the leading horologists of their day.

Thomas Tompion (1639-1713)

Thomas Tompion, (Fig. 9-1) is thought of as the greatest of English clockmakers. He has been called both "The Father of English Clockmaking" and "The Father of English Watchmaking," and when one looks at the immense diversity and on occasions complexity of his work, coupled with the extremely high standards which he kept, it will be appreciated just how justified that opinion is. It would not be too much to say that he raised the quality of English clockmaking to levels never previously attained, and it was he and a handful of other makers who gave England her pre-eminent position in the horological world, which was to last for seventy to eighty years. He was also an eminent scientific instrument maker.

Fig. 9-1. **Thomas Tompion 1639-1713.**

Tompion, the son of a blacksmith, was born at Ickwell Green near Northill, Bedfordshire in July 1639. A bell cast for St. Laurence Church at Willington near Northill is signed Thomas Tompion Fecit 1671. It may be that he served his apprenticeship as a clockmaker and worked in that part of the country prior to moving to London. However, it is more likely that he spent his early years working with his father as a blacksmith, and indeed the family forge still exists and bears a plaque commemorating its historic connection with Thomas Tompion. It must also be remembered that at that time many of the turret clocks, then called "Great Clocks", were still being made by blacksmiths and that when Tompion became a Free Brother of the Clockmakers' Company it was as a "Great Clockmaker."

An interesting point mentioned by Symonds in his book on Thomas Tompion[1] was that when his father, also Thomas, died in 1665, he left the tools of his workshop, not to Thomas, as might be expected, but to his younger brother, James, which would seem to imply that Thomas had already by then given up his work in the smithy, presumably in favor of clockmaking; however where and for whom he worked is unknown. The only certainty is that by the time he came to London at the age of thirty-two he was already a highly skilled and competent clockmaker with the confidence and knowledge to tackle any problem, be it a scientific instrument or complex clock. In July 1671, he paid a search fee to the Clockmakers' Company and was made a Free Brother on September 4th of that year, a significant time roughly corresponding with the invention of the anchor escapement and the introduction of the Royal or seconds' beating pendulum.

In 1674 Tompion moved to Water Lane at the sign of "The Dial and Three Crowns" where he was to continue his ever-expanding business for the rest of his life. Tompion was in his day a very highly respected figure, associating with members of the court and leading scientists, for whom he sometimes made astronomical and other scientific instruments. The poet Matthew Prior[2] (1666-1721) wrote, "Remember that Tompion was a Farrier and began his great knowledge in the Equation of Time by regulating the wheels of a common jack to roast meat".

The rise to fame of Tompion can only be described as meteoric. In the same year that he moved into Water Lane, he met Dr. Robert Hooke who commissioned a quadrant from him for the Royal Society. Hooke was the leading physicist and mathematician of the day; a man of great intellect and undoubtedly an inventive genius. It was his association and friendship which was to be the starting point for Tompion's rise to fame. It is also from his Diaries[2,3] that we learn much about Tompion's work.

In his preface to *The Artificial Clockmaker*[4] William Derham states, *In the History of Modern Inventions, I have had (among some others) the assistance chiefly of the ingenious Dr. Hooke and Mr. Tompion. The former being the Author of some, and well acquainted with others, of the Mechanical Inventions of that fertile Reign of King Charles the II and the latter actually concerned in all, or most of the late inventions in Clock-work, by means of his famed skill in that, and other Mechanick operations.* This gives some idea of Tompion's dominant position in English horology in the last quarter of the seventeenth century.

In 1675 Hooke asked for Tompion's help in proving that he had invented the spiral balance spring for watches prior to Huygens. The result of this was that Tompion was the first English maker to apply the spiral balance spring to a watch, making these more accurate than any produced by his competitors. Indeed, such was the demand for his watches throughout Europe that during his lifetime he, or to be more accurate, he and his employees, produced around 5,500. Charges for these are quoted as silver cased £11, gold cased £23, and gold repeater £70, a very large sum of money in those days particularly when one bears in mind that in 1675 Tompion bought his new premises on the corner of Water Lane and Fleet Street for £80.

In 1676, only two years after Tompion had moved into Water Lane, he was commissioned by Sir Jonas Moore to provide two clocks for the new observatory being built at Greenwich which Flamsteed was to run. The purpose of this observatory was to assist in navigation at sea by improving current knowledge of the fixed stars and the passage of the moon amongst them.

The concept of the clocks was, for their time, breathtaking. The dials were designed to appear in the room with the 13-foot, two-seconds'-beating pendulums suspended above them. Moreover, they were of year duration. This aspect of Tompion's life and work has already been dealt with in some detail earlier in the book and will not be reopened here.

Although Tompion produced a standard series of bracket and longcase clocks, they were nearly always made to a higher standard than other makers, his repeating work, seen here on a longcase clock (Fig. 9-2) was made to his own design, with repeating levers on either side of the backplate, and was undoubtedly the most reliable ever made. The dial layout and movement on an early longcase by him (Fig. 9-3), now in the British Museum, shows considerable originality and it is only towards the end of the century that his dials and movements became relatively standardized (Fig. 9-4). He was always willing to undertake special commissions, however complex, some of which were way beyond the abilities of most of his contemporaries and illustrates the great depth of his knowledge and understanding of mathematics and astronomy. One of the most remarkable of his achievements was a small and very beautiful spring or table clock that he made for William III which had a duration of one year. This is now known as "The Mostyn Tompion."

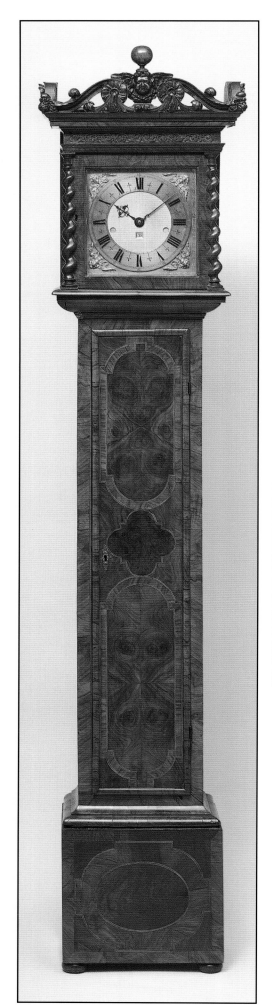

Figs. 9-2A, B, C. **Tompion No. 64. A walnut cased month going longcase clock with pull quarter repeating work, bolt and shutter maintaining power and calendar.** Unusually the five wheel going train is on the left and the repeating may still be operated when the clock is on silent. c. 1685.

Fig. 9-2C

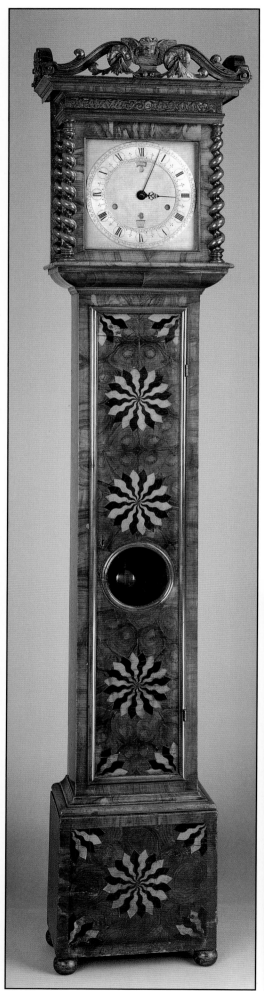

Figs. 9-3A, B, C. **Thomas Tompion. An un-numbered olive-wood and parquetry month duration calendar clock. c. 1680.** The dial layout of this clock is highly individual. The Roman hour numerals are ringed; the inner edge of the chapter ring has five minute and half hour marks and the outer edge of the chapter ring has every minute individually marked and also has 15 second divisions. Below 12 o'clock are apertures for the day of the week and the ruling deity and above 6 o'clock the day of the month and the month of the year with the number of days in that month.

The movement is somewhat unusual in that the going train is on the left and there is no center wheel as such, the drive to the motion work being taken from the pinion and wheel mounted on the extended second wheel arbor. Formerly in the Ilbert Collection, now in the British Museum.

Fig. 9-4A, B, C. **Thomas Tompion No. 371. c. 1700.** A month duration mulberry wood veneered longcase clock with bolt and shutter maintaining power and count-wheel striking.

Besides clocks of long duration or with complex striking, he made his celebrated angle clock[5], (Fig. 2-8) now at the National Maritime Museum, another incorporating an astrolabe (Fig. 9-5) and several equation clocks (Figs. 9-6, 9-7). An interesting sidereal and meantime clock, originally signed Tompion and Banger but covered with a plaque signed just by Tompion and fitted with a dead beat escapement is seen in Fig. 9-8.

Fig. 9-5A, B, C. **Astrolabe Clock by Thomas Tompion, c. 1677.** It is clocks such as this that illustrate so well the meteoric progress of Tompion's career, bearing in mind that he was making clocks of this complexity only six years after he came to London. It was owned by the late Mr. Ernest Prestige and was presented to the Fitzwilliam Museum Cambridge in 1948.

The Movement. The movement is of month duration and has a very early form of anchor escapement with the escape wheel revolving in an anti-clockwise direction, possibly to allow for the unusual train. It is fitted with maintaining power but no shutters are provided.

As it would be impossible to wind the clock through the center of the dial, extensions have been fitted to the bottom of the plate on which are mounted separate winding arbors which are geared to the main arbors.

The Dial. The dial has a 24-hour (2 x I-XII) chapter ring indicating mean time with a straight steel hand for the minute and a graduated hand carrying the sun at its' tip for the hours. The chapter ring, two inner rings and the spandrels are made of silver.

Attached to the central ring is a moon hand which indicates lunar time on the chapter ring. Within the central ring is an aperture to show moon phases and engraved on it are the age of the moon and the time of High Water at London Bridge.

The rotating fret within the chapter ring has engraved on its' rim a year calendar, the zodiacal calendar and the position of the sun in the ecliptic. Behind the fret may be seen the curved lower border of the perceived horizon and engraved above this is a scale of azimuth lines.

The space immediately below the horizon colored light blue represents twilight and below this again is a dark blue crescent the edge of which represents the crepuscular line. When the sun drops below this there is total darkness.

Marked on the fret are a number of stars and thus as it rotates it can be seen when these cross the horizon and become visible. Their position can then be read off during the night against the azimuth and altitude scales engraved on the plate.

The Case. The case is decorated with olivewood oyster veneers set in panels and with delicate floral marquetry. The Fitzwilliam Museum, Cambridge.

Fig. 9-6. **The Drayton Tompion.** This clock, so called because it was in Drayton House, Northamptonshire for some 200 years, now resides in the Fitzwilliam Museum, Cambridge.

It is of year duration and has an 80-pound driving weight. The pendulum bob, visible through the decorative aperture in the lower part of the door, is backed by an engraved and silvered brass plate marked out in degrees to give the amplitude of swing of the pendulum. An unusual feature is two movable pointers that can be used to note the maximum arc on either side. The very substantial movement has the train reversed with the barrel at the top and is provided with maintaining power.

The dial, apparently square, actually extends up to carry the dial for the year calendar which may be seen through an aperture when the hood is removed. There is a cutout in the center of the dial for the days of the week and the ruling deity.

The chapter ring, which shows mean time, is a 24-hour one graduated I - XII and 0-60 twice. The minute hand thus completes one revolution every two hours. A further ring immediately outside the main chapter ring gives apparent solar time. This is made possible by rotating it backwards or forwards throughout the year, relative to the main ring. The control for this is a roller resting on the equation kidney that is mounted behind the year wheel in the arch, and is connected to the wheel that moves the apparent solar chapter ring via a rod and rack. Fitzwilliam Museum, Cambridge.

Left & above:
Figs. 9-7A, B. **Year Duration walnut cased Equation Clock by Thomas Tompion.** This clock was made by Tompion for William III, his best patron, towards the end of the 17th century. The beautifully executed timepiece movement, i.e. without striking, has an inverted train with the pallets of the anchor escapement mounted upside down and the pendulum swinging in front of the 60 pounds weight.

It is fitted with a 24 hour dial with a fixed chapter ring inscribed Equal Time, with I - XII twice for the hours, and delicate Arabic numerals marked 0 - 60 twice for the minutes.

Outside this is a further ring marked in minutes on which is engraved Apparent Time (apparent solar time) which moves back and forth throughout the year enabling not only apparent solar and mean solar (clock time) to be read off against the minute hand, but also the Equation of Time to be calculated.

There is an arch to the dial, probably the first occasion on which this feature was used, which contains two apertures, the upper one showing the signs of the zodiac and the sun's position in the ecliptic, and the lower one with a year calendar. Mounted behind this may be seen the equation cam. The winding square is situated below 12 o'clock and is protected by the shutter of the bolt and shutter maintaining power that is activated by a lever on the right hand side of the dial. To reduce friction, the minute hand is counterbalanced and the seconds' hand omitted. *Reproduced by gracious permission of Her Majesty The Queen.*

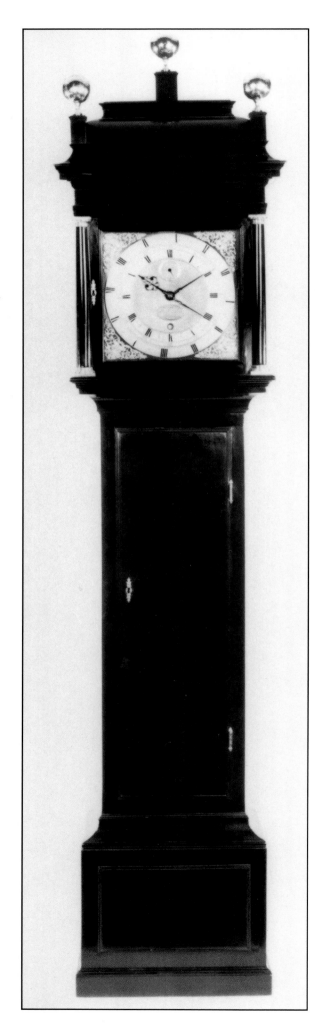

Figs. 9-8A-D. **Thomas Tompion Sidereal and Mean Time Clock, No. 483.** A rare and important sidereal and mean time month duration regulator signed Thomas Tompion London on an oval silvered plaque beneath which are the joint signatures of Thomas Tompion and Edward Banger, London. The concealing of Banger's name would seem to indicate a date of manufacture for the clock of 1708-1709 as it was in 1708 that Banger left Tompion's employ.

The 12" square dial has a finely matted center with a seconds' ring below 12 o'clock and a winding hole above 6 o'clock with a shutter connected to the maintaining power which is actuated by a lever situated just below 2 o'clock.

There are two chapter rings, a conventional inner one with half hour markings and quartering on the inside and an outer one which has minutes shown on its inner aspect and rotates twice a year.

As presently set up, the pendulum beats mean solar seconds and thus shows mean time on the inner ring and sidereal time on the outer rotating ring. To read sidereal time is, as described by Neilson[11] a relatively complex affair involving the subtraction from the meantime shown of the difference, counted anticlockwise, between XII on the rotating and fixed rings. Had the ring rotated anticlockwise, sidereal time could have been read off directly.

In view of the foregoing it would seem more likely, as Todd[12] suggests that the clock originally beat sidereal seconds, particularly as the outer ring rotates twice in 366¼ not 365¼ days. If this supposition is correct then the fixed chapter and seconds' ring will show sidereal time and the hour and minute hands, if extended, would show mean time on the outer rotating ring.

The function of the gilt minute hand is also a matter for some debate. Neilson suggests that it could have been used as a manually set equation hand. However another suggestion is that it may have been intended to indicate sidereal time for a given longitude East or West of the clock's location. The adjustment allowed via the curved slot in the gilt hands boss would encompass any two longitudes as far apart as Greenwich and Plymouth.

The single train movement has six latched plate pillars, four feet for the separate dial center, and a further four for the outer dial plate which is supported by extensions from the front plate of the movement. In this picture may also be seen the drive to the

outer chapter ring terminating in a spiral at its periphery, the four rollers on which the chapter ring runs, the shutter work and the 288 tooth hour wheel which meshes with the 24 leaf pinion of the 144 tooth minute wheel. This runs on an arbor that extends through the plates and then carries a 288 tooth first wheel of the differential gearing that in turn engages a second wheel of 144 teeth. At the bottom of this wheel arbor a worm engages a 15 leaf pinion at the bottom of the vertical arbor which runs up to drive the outer chapter ring.

The brass rod pendulum which is suspended from the backplate has a calibrated rating nut and is provided with a beat scale. A securing bracket is used to fasten the movement to the backboard.

The escapement is, in many ways, the most interesting feature of the clock in that it has been fitted with what is known as a Graham dead beat escapement. However, there can be no doubt that the clock was made several years before Graham is attributed with having invented the dead beat. Various explanations have been put forward for this, the principal one being that Graham changed the escapement, probably after Tompion's death. However the report of the Research Laboratory for Archaeology and the History of Art of Oxford University suggests that the escapement fitted was original to the movement. If this is so, then either Tompion invented this form of dead beat or Graham did so whilst still in Tompion's employ and before he became a partner; however it must be acknowledged that all that was proved for certain by the Research Laboratory was that the metal the escape wheel was made of could be the same age as the clock.

In about 1680 Tompion started numbering his clocks and watches, a practice which he continued until his death and which George Graham carried on after him. A list of the numbered clocks and watches is contained in *Clocks and Their Values* by Donald de Carle, first published by N.A.G. Press in 1968.

In 1696 George Graham joined Tompion at "The Dial and Three Crowns" and ten years later married his niece Elizabeth. Edward Banger also worked with Tompion, and between 1701-1708 the products were signed in their joint names, but in 1708 Banger left Tompion for reasons unknown and from 1710 to 1713, when Tompion died, the clocks and watches were signed "Tompion & Graham."

Tompion was buried in Westminster Abbey where his gravestone may still be seen. He left the business and most of his estate to George Graham and his wife.

For a fascinating and superbly researched and documented account of this maker, the reader can do no better than refer to R.W. Symonds *Thomas Tompion, His Life and Work*, B.T. Batsford Ltd., London and New York. Within a couple of years this will be joined by another book on Tompion, written by Jeremy Evans of the British Museum to which many years of dedicated study has been devoted.

George Graham (1673-1751)

George Graham (Fig. 9-9) was born in 1673, this fact being confirmed by the placing on view at the General Register Office, East Hall, Somerset House, of documents to that effect, c. 1951 but no one knows for certain the exact place where he was born; however he undoubtedly came from a Cumbrian background and this has been excellently researched by John Penfold.[6] A very brief summary of the complex trail he unfolds is as follows:

George Graham Senior, who became a widower after 1662, farmed Horsegills in the Kirklington area of Cumberland and later took into employment and subsequently married, a Scots woman Isobel by whom he had three children, George Junior, born late in 1673 or early 1674, possibly out of wedlock; John 1675 and Isabel 1677.

Fig. 9-9. **George Graham 1673 - 1751.**

George Senior died in 1679, by which time he had become an ardent Quaker. His son George, who would have been about six at the time, was brought up by his half brother from his father's previous marriage at Sikeside in Kirklington, where he seems to have had a reasonable education. In 1688 Graham went to London accompanied by Richard and Mary Graham, children of his father's first marriage; however unlike them he appears never to have become a Quaker.

He was apprenticed to Henry Aske on 2nd July 1688 but did not live with him during the second part of his apprenticeship. At this time he had come to know Thomas Tompion[7] and his family. Graham gained his freedom of the Clockmakers' Company on 30th Sept. 1695 and joined Tompion's household at the Dial and Three Crowns on the corner of Fleet Street and Water Lane at the end of 1696, where he was to stay for some seventeen years, subsequently marrying Elizabeth, the daughter of Tompion's brother James.

Graham, whose career is outlined by Atkins and Overall[8], joined Tompion in his business, becoming a partner in 1711, and when Tompion died in November 1713 succeeded him. He became a member of the Court of Assistants of the Clockmakers' Company in 1716 rising to Warden from 1719-1721 and Master in 1722. Graham was considered the greatest horologist of his day and also a very fine mathematical and astronomical instrument-maker.

He is particularly well known for his work on compensated pendulums, investigating the co-efficients of expansion of various metals and finally designing and producing the mercury compensated pendulum. He is also remembered for the form of dead beat escapement he produced which became almost universally adopted. Both these matters have already been discussed at some length earlier in the book.

In 1724 he greatly improved Tompion's horizontal escapement which was to be adopted by most of the great clockmakers of his day for their watches. At the same time he constructed a great mural arch at the Royal Observatory, Greenwich and devised a sector that enabled Dr. Bradley to discover two new motions in the fixed stars.

Graham was the first person in England to produce a planetarium. This was for Lord Orrery, whose name was subsequently given to these machines. He also supplied the French with the instruments needed when they were sent to the north to make observations for assessing the shape of the earth. This was to be the pattern of things to come, England becoming the center for the supply of astronomical and other instruments for the rest of the eighteenth century. Indeed Graham was far better known throughout Europe for his scientific instruments than his clocks and watches.

By 1720 he had become a fellow of the Royal Society where he gave lectures on his various discoveries. In 1722 he became a member of the Council and served on it on numerous occasions.

His pattern of clock and watch production in many ways followed that of Tompion. His table clocks were similar to Tompion's in their basic design and continued on with relatively little change until his death on 20th November 1751 and the longcase clocks, which were always of fine quality, varied little throughout his period of manufacture; however it is likely that, as with Tompion, his primary source of income was his watches.

Graham designed and produced a type of longcase regulator which was used in several observatories (Figs. 9-10, 11, 12), including the Royal Observatory at Greenwich (Fig. 9-13) who purchased two in the 1720's and one in 1750. It is thought that most of these were made for him by Shelton who continued their manufacture, in virtually unaltered form, after Graham's death. They were usually of month duration, had substantial plates with champhered top corners, six heavy latched baluster pillars, bolt and shutter maintaining power, stop-work and a screwed on dial, usually 10-10.5" square, with an aperture for the hours, each of the four dial feet being held in place by three screws.

Fig. 9-10. **George Graham, London. No. 756. c. 1745.** This movement, which is contained in an oak case, is typical of those used by Graham. It is of month duration and has bolt and shutter maintaining power and stop-work that may be seen to the left. There are six substantial latched plate pillars and four heavy screwed dial feet. The delicately laid out silvered dial has a very narrow chapter ring with Arabic numerals and a similar seconds' ring. There are apertures for both the hours and the date.

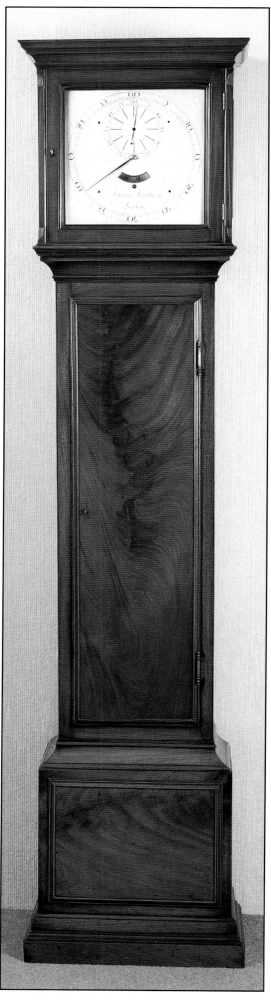

Figs. 9-11A, B, C. **George Graham, London. c. 1750.** The movement of this regulator, which is contained in a fine mahogany case, is similar to that seen in Fig. 9-10 but the wheelwork in the train has five crossings instead of six and the pallets, which are original, have been jeweled at a later date. Beat regulation and a beat plate are provided and the gridiron pendulum is suspended from a cock bolted to the backboard. The movement is protected by a wooden dust cover and the seconds' ring has bold observatory bars.

Fig. 9-12. **George Graham No. 767. c. 1745**. The movement and dial of this regulator is similar to that seen in Fig. 9-11 but it is contained in an oak case of simple pleasing proportions.

It is likely that the choice of timber for the case was, at least sometimes, dictated on economic grounds; for instance in the main entrance hall to one of the important country houses is a regulator with a fine mahogany case. Below stairs is another regulator with a movement of equal quality, but in a simple oak case.

Fig. 9-13A, B. **George Graham, London. Unnumbered but known as No. 3.** This month duration regulator was bought by Bradley for the Royal Observatory, Greenwich in 1750 and is still there today. Its history is given in some detail as it emphasizes that regulators were, for astronomers, working instruments, and all they were concerned about was their performance and reliability, not in maintaining their originality. It started life with steel pallets and a gridiron pendulum and was fixed to the south wall of the transit room where it was used as the sidereal clock for Bradley's 8ft. transit instrument, which defined the Greenwich Meridian from 1750-1816. It was subsequently adjusted to Solar Mean Time and used for dropping the time-ball from 1833-1924 when it was retired from active service, being moved first to Herstmonceux and then to Cambridge and finally back to the Observatory at Greenwich.

In 1771 Arnold fitted ruby pallets and in 1779 steel; Harrison's maintaining power was fitted and the pivot holes for the escape wheel and pallet arbors jeweled.

In 1780 it was moved to a stone pier near the transit instruments and the pendulum mounted independent of the clock case.

In July 1789 a solid brass bob supported at its center was substituted for the lead filled one and a small subsidiary bob was fitted for fine regulation. At the same time a special gridiron pendulum was made and the hour circle on the dial changed by Larcum Kendall from 12 to 24 hours. Three years later the motion work for the hour hand was reduced from three wheels to one and the single cross stay on the pendulum was replaced by Earnshaw with three.

By 1828 it was being used for pendulum experiments by Sabine and Kater and in 1836 the duration was changed to eight day and a mercurial pendulum fitted. National Maritime Museum.

Simple but well constructed rectangular cases were employed, which was to set the pattern for much of the century. Some were of oak (Fig. 9-12) others mahogany (Fig. 9-14), the choice probably depending on the situation in which the regulator was employed.

Although Graham devised the mercurial pendulum and got good results with it, he generally employed the gridiron, probably because it was far easier to transport, bearing in mind Graham's extensive export market, and less likely to be damaged or disturbed.

Besides his standard regulators he also made a small number of greater complexity such as that in the British Museum (Fig. 9-15) which gives mean and solar time, the Equation of Time, the time of sunrise and the sun's position in the zodiac.

Fig. 9-14. **A month duration mahogany longcase regulator made by Graham for the King of Spain.**
Patrimonio Nacional, Madrid.

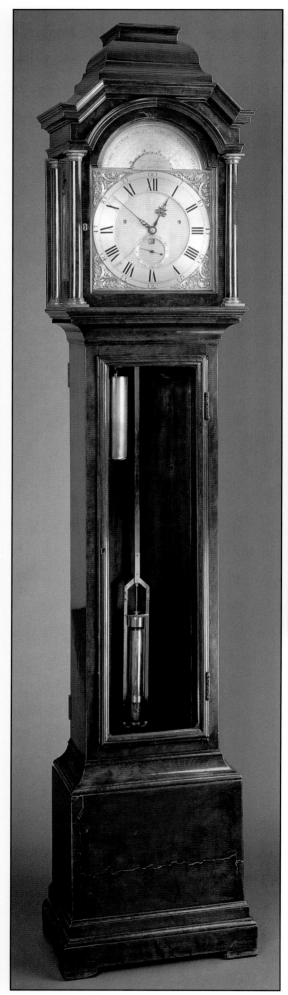

Figs. 9-15A, B, C. **George Graham, London. A month duration longcase regulator showing mean and solar time, formerly in the Ilbert Collection and now in the British Museum.** It has a dead beat escapement, Graham's mercurial pendulum and bolt and shutter maintaining power.

The blued steel hand indicates Mean (Solar) Time and the gilt hand Apparent Solar Time. There is an aperture for the date.

The engraved and silvered dial in the arch rotates once a year and the scales are read against the index in the center of the arch. These show, reading upwards: the time of sunrise; the sun's position in the zodiac with each sign subdivided into degrees; the sun's declination; the Equation of Time; the month and the date.
British Museum, London.

When Graham took over Tompion's business in 1713 he remained in Fleet Street at the corner of Water Lane, but in 1720 moved to "a new house next door to the Globe and Marlborough's Head Tavern."[9] This was 135 Fleet Street, where he was to remain until he died.

Graham shared the building with Thomas Wright for nine to twelve years and later other people. Wright was an instrument maker who became famous for his Orreries and mathematical instruments and was maker to the Prince of Wales. It would seem very likely that he and Graham shared their common interests.

Graham was succeeded by his foreman Samuel Barkley (or Barclay) who went into partnership with Thomas Colley. By 1754 Barclay had retired or died and Colley was to continue on his own until he too died in 1771.

Interestingly Mudge, who was apprenticed to Graham on 4th May 1730; freed of the Clockmakers' Company in 1738 and subsequently became his journeyman, set up at 151 Fleet Street in 1750 where he was to remain until 1772 when he moved to Devon. Dutton, who was also apprenticed to Graham (in 1738) then continued at this address until 15th January 1773 when a fire consumed much of the premises. In 1774 he moved to new premises at 148 Fleet Street.

A mark of the esteem in which George Graham was held, was that he was buried alongside Tompion in Westminster

Abbey, and moreover, that the pallbearers at his funeral included many eminent scientists.

Following Graham's death the following obituary appeared in the London Daily Advertiser which gives an idea of the high regard he was held in:

Saturday evening died suddenly at His house in Fleet Street Mr. George Graham not less known in the Learned World than in that branch of Business to which for many years so successfully applied himself as by his uncommon Ingenuity to have acquired the Reputation of being the best Watchmaker in Europe. He was many years Fellow and one of the Council of the Royal Society. His Apparatus made for measuring a Degree of the Meridian in the Polar Circle is greatly esteemed among the Literati: as are also his many curious Instruments for Astronomical Observations. He lived beloved and died universally lamented.

When Graham took over Tompion's business in 1713 William Webster placed the following advertisement in the London Gazette to try and secure some of the goodwill:

LONDON GAZETTE, Nov. 24-28, 1713. ON the 20th instant. Mr. Thomas Tompion noted for making of all sorts of the best Clocks and Watches, departed this Life. This is to certify to all Persons of whatever Quality or Distinction, That WILLIAM WEBSTER, at the Dial and 3 Crowns in Exchange-Ally, London, served his Apprenticeship, and liv'd as a Journey-man a considerable time, with the said Mr. Tompion, and by his Industry and Care, is fully acquainted with his Secrets in the said Art.

Graham was then forced to retaliate with an advertisement in The Englishman.

A similar state of affairs was to occur just two days after Graham died; when the following advertisement by Thomas Mudge appeared in The General Advertiser, The London Daily Advertiser, The Whitehall Evening Post and the General Evening Post:

THOMAS MUDGE, WATCHMAKER, late Apprentice to Mr. Graham deceased, carries on Business in the same Manner Mr. Graham did, at the Dial and One Crown, opposite the Bolt and Tun, Fleet Street.

This forced Barkley to respond as quoted below and follow it up with other advertisements:

To the Nobility Gentry etc., Customers of my late Master Mr. G. Graham, Clock and Watchmaker. Now that my Master is interr'd I think myself obliged to acquaint his Customers that the business of the Shop is carried on by me, in partnership with Mr. Th. Colley (we being the executors of Mr. Graham's will) that as the Business has been for several Years conducted by me and under my Inspection so everything goes on in the same Train as before upon my Master's Principles and the same Curious Artists in all the Branches of the Business are retained by Your most obedient humble servant. Samuel Barkley, Foreman to the late Mr. Graham.

An excellent article written by Charles Aked[10] titled "Three letters from George Graham" relates to experiments he instigated in Jamaica which were carried out by a Mr. Harris.

His first letter dated 15th July 1732 refers to the different time kept by a pendulum of exactly the same length, in London and Jamaica. He noted that after allowing for the different temperature in the two places that it went two minutes, six seconds slower in London in a sidereal day, something that Sir Isaac Newton had postulated, because of the shape of the earth, which is slightly flattened at the poles.

His second and third letters dated July 22nd and 25th 1732, concern his use of two thermometers, one spirit and the other mercurial. Interestingly, he refers to the latter as having been used by him for at least twenty years. Thus as early as 1712 Graham must have been well aware of the expansion of mercury.

In his third letter Graham analyses the allowance he should make for the different temperatures in London and Jamaica, taking into account the day and night time temperature, and concludes that the maximum error they are likely to make is four to five seconds a day.

On the back of one of these letters is a corrected draft of a paper on Isochronal Pendulums which was published in *Philosophical Transactions* Vol. 38 p. 312. In this he expresses the hope that people who take pendulum clocks to different countries will record their time-keeping without altering the length of the pendulum and also recording the temperature and barometric pressure. He points out that a change in length of the pendulum of just 0.01" corresponds to eleven seconds a day and thus it would be very easy to calculate the length of the pendulum required to keep exact time, which would relate to the gravitational force at that point and thus its distance from the center of the earth.

Interestingly William Derham commented on these matters in 1698, first discussing the effects of running a pendulum in a vacuum and later the difference between the timekeeping at different parts of the world. His conclusion in 1735, the year that he died, which was published in *Philosophical Transactions* for the years 1735-1736 concludes: *All these experiments seem to concur in resolving the Phaenomenon of Pendulum Clocks going slower under the Aequator than in the latitudes from it. I leave it to the consideration of others, how far heat and Cold or the Rarity and Density of the Air are concerned in this Phaenomenon.*

At the end of the article containing Graham's letters, Aked's comments on James Bradley give us some insight into Graham's importance in the world of astronomy, mentioning instruments he supplied to Bradley, including a 24 ft. telescope, a 12-1/2 ft. zenith sector and an 8 ft. quadrant. Charles Aked's summary of Graham's achievements, well worth quoting, is given here. *Great though Graham's renown as a clockmaker may be, his achievements in the field of astronomical instruments and the importance of the results obtained through their use, far outweigh the importance of his contributions to timekeeping, even though his were the greatest of all the achievements in the mechanical horology applied to time measurement on land, at that time.*

References

[1] Symonds, R.W. *Thomas Tompion. His Life & Work*. B.T. Batsford Ltd. 1951.

[2] Prior, M. *Essays and Dialogues of the Dead*. Cambridge University Press.

[3] Robinson, A.W. & Adams, W. Editors. *The Diary of Robert Hooke 1672 - 1680*.

[4] Derham, W. *The Artificial Clockmaker*. James Knapton. London 1696.

[5] Howse, D. *The Tompion Clocks at Greenwich & the Dead Beat Escapement Part II*. Antiquarian Horology. March 1971. pp. 116, 117 and 122.

[6] Penfold, J. *The Cumbrian Background of George Graham, Clockmaker*. Antiquarian Horology. March 1974. pp. 600 - 613.

[7] Penfold, J. *The London background of George Graham*. Antiquarian Horology. Sept. 1983. pp. 272 - 280

[8] Atkins, S.E & Overall, W.H. *Some Account of the Worshipful Company of Clockmakers of the City of London*. Private Publication 1881. pp. 166 - 167.

[9] Millburn, J.R. *The Fleet Street Address of Graham and his Successors*. Antiquarian Horology. June 1973. pp. 299 - 301.

[10] Aked, C. *Three letters from George Graham*. Antiquarian Horology. Winter 1988. pp. 597 - 605.

[11] Neilson, M. *Important Sidereal Regulator by Thomas Tompion and Edward Banger, No. 483. c. 1709*. Antiquarian Horology Vol X, No. 2. pp. 214 -216.

[12] Todd, W. *Communication in Antiquarian Horology*. Vol. X No. 3. p. 366.

Chapter 10
John Harrison 1693-1776
by Derek Roberts

John Harrison (Fig. 10-1) is best remembered today as the man who eventually gained the government's prize of £20,000 by producing a clock which was sufficiently accurate for ships to determine their longitude at sea and enable them to navigate with safety; a feat which was to transform trading throughout the world.

Fig. 10-1. **Portrait of John Harrison engraved by Bernet Reading.**

John, one of a family of five, was born to Henry and Elizabeth Harrison of Wragby, Yorkshire in May 1693. Of his two younger brothers only James, eleven years his junior, also became a clockmaker. The Harrison family tree, based on the research of Colonel Humphrey Quill was published in 1966[1], and amendments and corrections by Andrew King, were included in *The Quest for Longitude*.[2]

Henry Harrison, a carpenter and well respected member of the local community, worked for Sir Rowland Winn, first at Nostell Priory and later at Barrow, near Barton on Humber. He would have taught John his skills as a carpenter and joiner and helped with his general education.

However, John Harrison obviously had a deeply enquiring mind and he was assisted in this respect by a local clergyman who, amongst other things, lent him a manuscript copy of Saunderson's lectures on mechanics and physics.[3] These he proceeded to copy in full, including the drawings. The knowledge he gained from this enabled him to make some money by land surveying and, quite probably, triggered off his interest in clocks.

By 1713, at the age of 20, John Harrison had made his first clock (Fig. 10-2), the movement of which is now in the collection of the Worshipful Company of Clockmakers. The case no longer exists, a state of affairs that also applies to his second clock made in 1715 (Fig. 10-3) which is now in the Science Museum. A third clock, dated 1717, in a later case, is preserved by the National Trust at Nostell Priory.

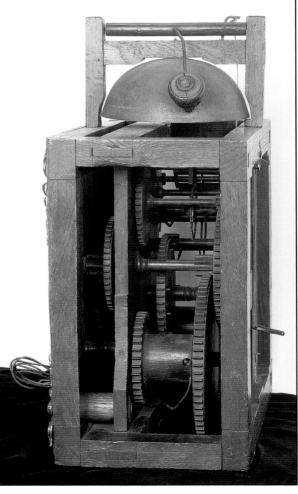

Fig. 10-2A, B, C. **The movement of John Harrison's first longcase clock, c.1713 made when he was 20, now with the Worshipful Company of Clockmakers.** The substantial joined oak frame, made out of 1" thick timber, measures 14.5" x 12." The only metal employed is steel for the pivots and brass bushes, the escape wheel, the pallets with crutch and the pins in the date wheel. It is signed and dated by him on the date wheel.

Above and following two pages:
Figs. 10-3A, B, C. **Harrison's second longcase clock, c. 1715 now in the Science Museum, London.** Seen here with and without the dial and also the wheelwork removed.
  The wooden dial plate, which is pinned to the frame, is gilded and has an applied brass chapter ring and spandrels, the lower two of which may be removed to give access to the winding wheels which employ reduction gearing. He often signed and dated his clocks on the date wheel.

Fig. 10-3B

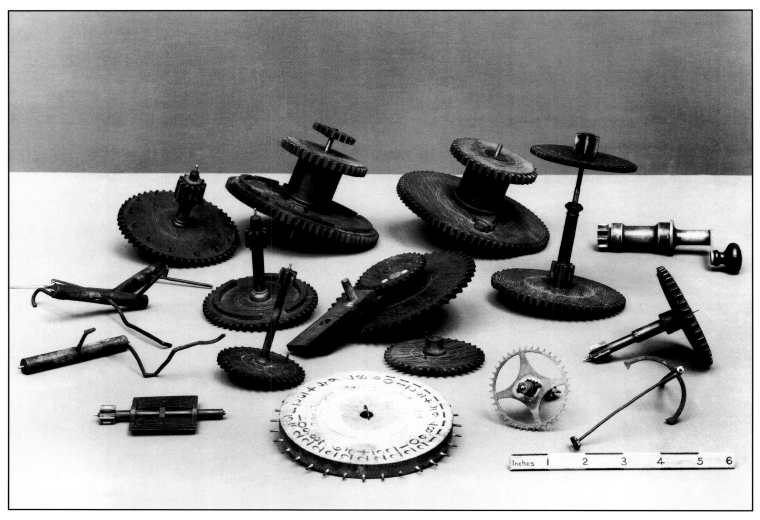

Fig. 10-3C

Because of John Harrison's training, although the clocks were of relatively conventional design, they were made almost entirely of wood with very substantial jointed oak frames just under 1" thick and approximately 14.5" x 12".

The wheelwork was of oak with the great wheels, some 7" in diameter, meshing with ten leaf pinions which, like all the other pinions and arbors, employed boxwood. The pivots were of steel running in brass brushes and the escape wheel was also of brass. The only other metal employed was for the steel pallets with crutch and the pins in the wooden date wheel.

The way in which Harrison constructed his wheels was to turn them out of oak, cut a slot into the center of the periphery and then insert the teeth into this slot in groups of three or four, with the grain running the length of the tooth. All three clocks made in 1717 employed an anchor escapement.

The dial, like the movement, was made of wood and then painted and gilded. Interestingly, the lower two spandrels were made removable (Figs. 10-3) to give access to winding gears on the barrels. The key is provided with a pinion on the end.

One of John Harrison's great strengths in his early days was that he lived in a relatively isolated area and had no formal training as a clockmaker, let alone having been apprenticed to one. He thus had no preconceived ideas as to how a clock should be made other than by looking at those he saw or repaired. This gave him a freedom of thought denied to others that was to stay with him for his entire life.

This came to the fore when in the early 1720s he was asked by Sir Charles Pelham to supply a clock for his new stable block at Brocklesby Park (Fig. 10-4). Colonel Quill[4], who was the first to realize the significance of this clock, examined and documented it. He suggested a date of manufacture of circa 1727; however, Andrew King[5] feels that this date is unreliable and that a more realistic one is 1722, about when the stable block was completed. It is at this time that James, John's younger brother, who at the age of eighteen was a skilled carpenter, probably joined him and assisted in the construction of the Brocklesby Park clock (Fig. 10-5), which would account for the fact that it was just signed Harrison. However, it is very unlikely that James had any hand in its design or even that of the later precision clocks that are signed and probably made by him. It is known that he came down to London in 1737-1738 but, thereafter, returned to Barrow to continue his career as a carpenter, joiner and bell hanger.

Fig. 10-4. **The Turret Clock at Brocklesby Park.** Harrison's double dialed clock, which was commissioned by Sir Charles Pelham c.1722, sitting on top of the roof of the stable block at Brocklesby Park, now the seat of the Earl of Yarborough, a title granted to the Pelham family in 1837.

Fig. 10-5A-K. **The Movement of the Brocklesby Park Clock.** This is thought to have been made circa 1722 by John Harrison and his brother for Sir Charles Pelham's new stable block. As originally conceived it bore a close family resemblance to his earlier clocks with anchor escapement, but used lignum vitae for the bushes. A knife edge suspension resting on glass was fitted but whether originally or at a later date is not known. Because Harrison ran into difficulty with the clock he designed and fitted his famous grasshopper escapement which, because there is no sliding friction on the pallets, does not require lubrication. He fitted roller pinions, and cycloidal cheeks of rudimentary form were also added.

Fig. 10-5A. Harrison's clock, firmly supported by a heavy oak beam and protected by its wooden case, the front doors of which open up to reveal the movement.

Fig. 10-5B. An overall view of the movement, still in remarkably good condition after 280 years and requiring very little attention. Note the extensive use of lignum vitae bushes.

Fig. 10-5C. The original winding key, which has been in retirement for many years having been replaced by a crank key. The mortise and tenon joint, top right, is still as tight as the day it was made, thanks largely to the relatively cold and damp conditions in which the clock is kept.

Fig. 10-5D. The oak handset dial to the rear of the clock, still with traces of its original decoration. The oak pegs retaining all the joints may be clearly seen.

Prior to the Brocklesby Park clock John Harrison had used relatively conventional clock designs, albeit with materials seldom employed by clockmakers in this country. However, with this clock he broke new ground, both so far as the original design was concerned and also the modifications that were required when problems occurred. It is these innovations that were incorporated in much of his later work, including his regulators and sea clocks.

One of the innovations incorporated in this clock was the use of lignum vitae, an extremely dense wood heavily impregnated with natural oils which never dry out, thus eliminating the need of oil for lubrication, which slowly congeals, increasing friction, altering the power coming through the train and thus the arc of swing of the pendulum and the timekeeping. A further advantage of the use of lignum vitae for the bushes (Fig. 10-5G) was that brass could then be used for the pivots instead of steel, thus eliminating the use of ferrous metals in the clock with their liability to corrosion, particularly in the hostile setting of a turret clock.

It is thought that in the first instance an anchor escapement with equal lift and thus impulse was fitted. To reduce friction on the escapement pivots to a minimum, a knife edge suspension resting on glass (Fig. 10-5F) was employed although whether initially or at a later date is not certain.

The fact that the clock proved unreliable could be called a blessing in disguise as it forced Harrison to design and fit various modifications which were to stand him in good stead when he produced his precision pendulum and later his sea clocks. These were as follows:

1. The grasshopper escapement (Fig. 10-5E), so called because of its fascinating action which is described in Chapter 6. The principal advantage of this is that there is no sliding friction at the pallet faces and, therefore, no lubrication is required. Proof of this is the fact that one of his sea clocks in the National Maritime Museum has now been running continuously without any servicing for some thirty-five years. There is also minimal friction at the pivots.
2. The use of roller pinions on the escape wheel arbor that consisted of lignum vitae rollers on brass wire. These may be seen best in what is believed to be his last clock, now owned by the Royal Astronomical Society (Fig. 10-11).
3. The employment of anti-friction rollers, in this instance to support the escape wheel arbor.

The result of all these modifications was that friction decreased dramatically and thus the arc of swing of the pendulum greatly increased. To reduce this to acceptable levels wind vanes were fitted to the bob of the conventional pendulum (Fig. 10-5J) which had a brass rod and brass suspension spring. A good example of Harrison's attention to detail is the use of lignum vitae inserts neatly dovetailed into the crutch to keep friction, which often occurs at this point, to a minimum (Fig. 10-5K). With all the hard won knowledge gained with the Brocklesby Park clock Harrison was now in a position to design his first precision pendulum clock.

Fig. 10-5E. Harrison's grasshopper escapement and below it the brass anti-friction roller supporting the escape wheel arbor.

Fig. 10-5F. The rear brass knife edge resting on a glass plate on which the pallets pivot.

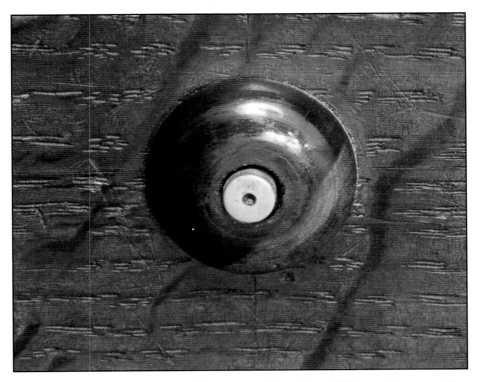

Fig. 10-5G. A lignum vitae bush with brass pivot. It is well over 100 years since several were replaced, but there is very little wear. Although one is now cracked and tends to work its way out every two years, it has not been replaced, just pushed back into position.

Fig. 10-5H. A close-up of one of the great wheels showing the very limited wear which has taken place in some 280 years.

Fig. 10-5J. The lead pendulum bob and the vanes which had to be added to it to reduce the increased arc of swing following the fitting of the grasshopper escapement. 10-5K

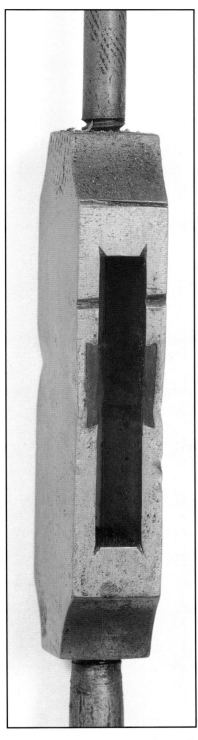

Fig. 10-5K. The brass pendulum rod and crutch with lignum vitae inserts to reduce friction to a minimum at this critical point.

Fig. 10-6. **This regulator, now in the Museum of Science & Industry, Chicago, and formerly with the Time Museum, was made circa 1725 - 26.** The case, with its Equation of Time table glued to the front door, is original. The escapement, gridiron pendulum and date wheel have been re-instated. Photo courtesy of the Time Museum.

By the end of 1726 Harrison had completed two regulators with gridiron pendulum and grasshopper escapement and in the following two years one (Fig. 10-6), and possibly two more. Stukeley, in his manuscript journal of 1728, refers to one of these, presumably the one Harrison retained for his own use (Fig. 10-7), as follows: *I saw his famous clock last winter at Mr George Graham's. The sweetness of the motion, the contrivances to take off friction, to defeat the lengthening and shortening of the pendulum; through heat or cold cannot be sufficiently admired.*

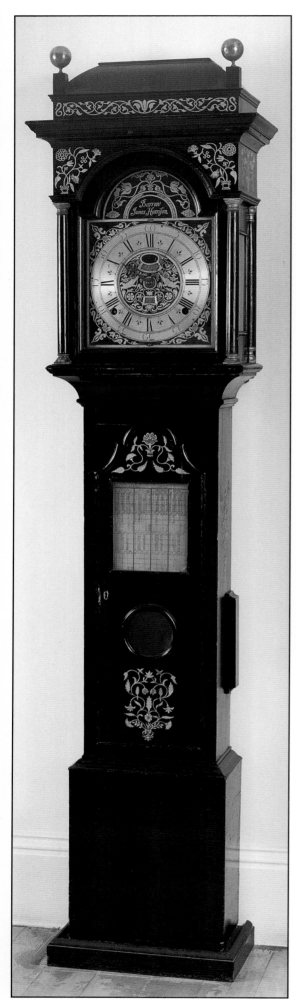

Pages 204-207:
Figs. 10-7A-G. **Harrison's 1728 regulator, now in the collection of the Worshipful Company of Clockmakers.** This regulator, which was kept by Harrison throughout his life, whilst still retaining the wooden frame, incorporates and improves on the modifications Harrison fitted to the Brocklesby Park clock. The design of the grasshopper escapement (Fig. 10-7F) was altered, using a single pivot point for both pallets. Oak wheels were employed, as on his earlier clocks, and also oak arbors with brass pivots to either end. Roller pinions were used throughout the train and lignum vitae employed for the pivot holes. Two antifriction rollers were fitted for the escape wheel arbor. The oak dial is painted black with gilt decoration and has an engraved brass chapter ring. The trunk door still retains Harrison's equation table. Probably the single most important feature of the design was the use of Harrison's gridiron pendulum, the one on this clock being the only gridiron by him known to still exist.

Fig. 10-7C

Fig. 10-7D

Fig. 10-7E

Fig 10-7F

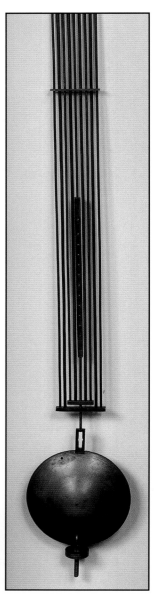

Fig. 10-7G

The records of The Worshipful Company of Clockmakers[6] go on to quote Dr. Hutton's statement that the timekeeper did not err by one second a month.

The accuracy of Harrison's regulators, and his assertion that they kept time to within one second a month, has long been a matter of debate with authorities such as Andrew King[5], Martin Burgess[7] and W S Laycock[8] basically defending this claim; however, research by Bateman and James[9] into the quality factor of the pendulums of four clocks by John Harrison casts doubt on this.

The "quality factor," abbreviated to Q, the relevance of which is currently under some debate, has come to the forefront in horology in recent years and has been applied elsewhere for much longer. It basically records the efficiency or, to put it another way, the rate of decay of an oscillating body, such as a pendulum, and is a numerical constant. This makes it easy to compare its efficiency with various other types of oscillator such as a balance wheel or even a quartz crystal. It takes no account of variable factors such as temperature and barometric pressure and thus when measuring it is assumed that these are allowed for.

Bateman[10] has shown that, at least in theory, the short term accuracy of a clock is directly related to its Q factor and in 1985 recorded the results of his and James' research into the pendulums of four of Harrison's clocks.[11] The average Q factor was found to be around 5,000, which is about half to one third of that of a typical observatory regulator and thus, on the basis of its Q value would probably not be able to match their nominal +0.7 seconds per day error; however W5, produced by Woodward (Chapter 39) has a relatively low Q and yet is a superb timekeeper.

M. K. Hobden[12] and D. Bateman[13] produced interesting comments on Bateman and James' research and it is helpful to read these at the same time as the original article.

Much has been written on the Q factor, some of which is of a completely analytical nature. Whilst this is of interest to those (and some others) who are still pursuing the ultimate precision pendulum clock, it is felt that it is beyond the remit of this part of the book to delve into this subject further.

Whereas on the Brocklesby Park clock a single roller was employed to support the escape wheel arbor (Fig. 10-5E), two were fitted on the regulators, a much more efficient arrangement, and roller pinions were used throughout the train. Probably the most interesting clock to have survived is that made in 1728 which was retained by Harrison throughout his life, together with one other, and is now with the Worshipful Company of Clockmakers (Fig.10-7). This incorporates Harrison's gridiron pendulum, perhaps, after his form of maintaining power, the best known and certainly the most universally adopted of his inventions, which makes use of the different coefficients of expansion of two metals, usually iron and brass, to cancel each other out and thus give a pendulum of invariable effective length. This came only a few years after Graham's invention of the mercury compensated pendulum and the principle was used on Harrison's sea clocks. It is interesting to compare Harrison's drawing of the pendulum in his 1730 manuscript and the pendulum from his 1728 regulator (Fig. 10-8) with that of the gridiron pendulums used in later years. The rods are much thinner, allowing far more rapid heat transference and thus response to changes in temperature. Although Harrison's original pendulum may not have looked so impressive as some later examples, particularly the French ones, it was probably far more efficient. He even made the brass rods thicker than the steel ones to allow for the differences in conductivity and specific heat of the two metals.

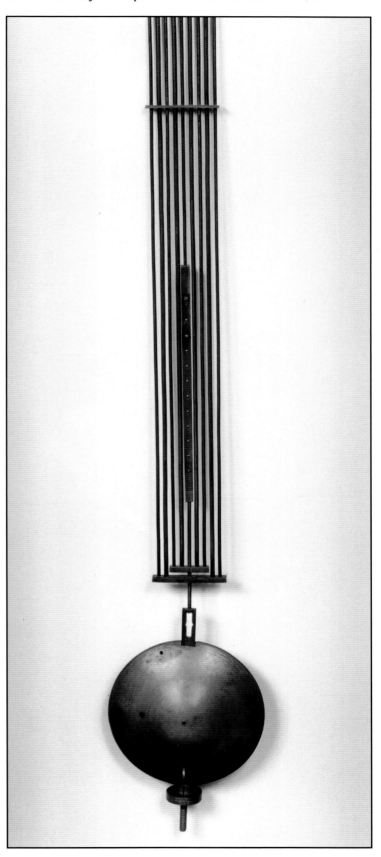

Fig. 10-8. **Harrison's gridiron pendulum which was fitted to the 1728 regulator.** Note the holes in the central rod so as to be able to adjust the compensation.

One of the big disadvantages of the conventional gridiron pendulum is that the lengths of the rods relative to each other cannot be varied without unpinning and re-pinning them. Harrison overcame this by incorporating a device to enable you to vary the length of the central, and possibly also other rods and thus the compensation. It is discussed and illustrated by W S Laycock[8] who describes it as breaking the central rod, which is steel, with an adjustable sandwich in which the meat is steel and the bread brass (Fig. 10-8).

A pin is passed through one of the holes and this unites them. He calculates that the shift of one hole by the pin would compensate for a temperature change of 13°, which would still allow the clock to stay within one second in 100 days. It is at this stage that we meet up with Harrison's quest for extreme accuracy, resulting in his claim of a deviation of no more than one second a month for his regulators. Interestingly, if one looks at the painting of Harrison by Thomas King (Fig. 10-9) carefully, a series of small holes may be seen in the outer rods, presumably to make it possible to adjust the compensation.

Fig. 10-9. **Oil painting of John Harrison by Thomas King**. If you look carefully you will see the perforations in the outer rods so that the compensation may be adjusted.

To adjust and check the efficiency of his gridiron pendulum he had to devise a method of testing them at various temperatures and this he describes in his 1730 manuscript, an extract from which is reproduced here:

But, my other way is the better part of the Completion: and that is the two Clocks plac'd one in one Room & the other in another, yet so, that I can stand in the Doorstead, & hear the beats of both the Pendulums, when the Clock Case heads are off, & before or after the hearing can see the seconds of one Clock, whilst another Person count the seconds of the other: by which Means I can have the difference of the Clocks to a small part of a second. And in very Cold and Frosty Weather I sometimes make one Room very warm, with a great Fire, whilst the other is very Cold. And again the Contrary. And sometimes the like in Summer by the Sun's Rays in at the Windows of one Room, & also a Fire, whilst the other is close shut up and Cool. Thus I prove the Operation of the Pendulum Wires...And to prove or adjust the Cycloid to Vibrations perform'd in different Arches as requir'd,...I cause the Pendulum to describe such by increasing & decreasing the draught of the Wheels, & that by adding to & taking from the Weight; by which I can make 8 or 10 Times more difference, than Nature ever will, & yet the effect be nearly the same...as if Nature it self had alter'd the Weight of the Air so much.

Harrison's signature at the end of this manuscript is shown in Fig. 10-10.

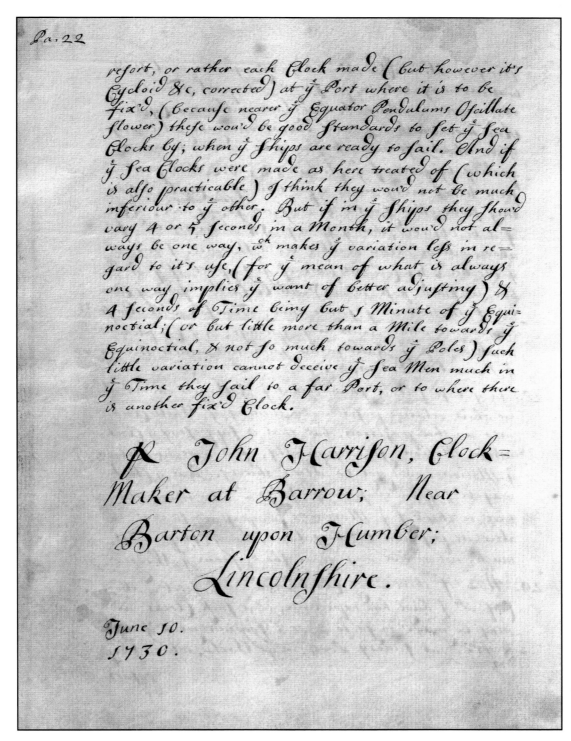

Fig. 10-10. **Harrison's signature at the end of his 1730 manuscript.**

To check the accuracy of his clock Harrison had to devise a precise means of recording the time. For this he used "Fixed Stars," sometimes called "Time Stars," situated so far away that for all practical purposes their positions may be considered as constant.

His method was to line up the east side of his neighbour's chimney with one of the uprights of his own window. An assistant would call out the seconds as the star approached and he would note the instant it disappeared behind the window frame.

As the eighteenth century progressed, astronomers refined their techniques, enabling them to record time to a much smaller margin.

Breguet[14], produced a star transit instrument (Fig. 28-14) on which the transit is recorded by observing the star through a tube with crossed wire sights. Rotating within the field of vision is a six armed spider, with each arm housing a small black disc at its tip. Every 0.1 second a disc crosses the observer's field and thus enables the transit to be checked to within 0.1 of a second.

It is interesting that even at this relatively early stage in precision timekeeping, (c. 1725-1730), not only was Harrison well aware of the nicety of providing the correct form of cycloidal cheek, but that he was also conscious of the importance of changes in barometric pressure[15] as evidenced by the following statement *"But a Pendulum moving in a Cycloid to regulate a clock true, which is not in vacuum, must not always keep its same length, unless the air gave always the same resistance, which agrees with both reason and experiment. Viz the pendulum must be shorter in warm rather than cold weather, which is contrary to the operation of the wire"*.

The research and thought which Harrison applied to his regulators from circa 1723-1724, when they were probably first conceived, until towards the end of the decade, when he became involved with his sea clocks, was immense, considering many facets such as circular deviation, the factors which affected it and how to correct it; all the effects of variations in barometric pressure and temperature, such as the change in the resistance of the air to the passage of the pendulum, variations in friction and the balancing of the power input to the arc of the pendulum, which he checked at various different temperatures.

It is beyond the scope of this book, and indeed of the author, to go into these matters in any detail but those who are seriously interested in Harrison's investigations into precision pendulum clocks can do no better than read Laycock's publication on *The Lost Science of John "Longitude" Harrison*[8] and Martin Burgess' stimulating and painstaking analysis of his research[7] together with Andrew King's contribution to *The Quest for Longitude*[5] edited by William J. H. Andrewes.

Having assimilated these, they may well reach the conclusion that Harrison's claim that his regulators kept time to within a second a month may well have been a valid one, which is reinforced by the performance figures so far available of the Gurney Clock[7], based on Harrison's pendulum technology, which was made by Martin Burgess.

In conclusion, we can do no better than quote from Harrison's 1730 manuscript as follows:

Some Years ago I made several alterations in order to render the Motion of Clocks more exact than heretofore, but when I came to try them by strict observation as below, I judg'd the best performance of the best Pendulum Clock I ever saw, made, or heard of, to be incapable of this Matter, wou'd it go as well in a Ship at Sea in any part of the World, as in any one fix'd place on the Land. Yet from several observations, I still endeavour'd to make farther Corrections in this Motion; and in these 3 last Years have brought a Clock to go nearer the truth, than can be well imagin'd, considering the vast Number of seconds of Time there is in a Month, in which space of time, it does not vary above one second, & that mostly the way I expect: so that I am sure I can bring it to the Nicety of 2 or 3 Seconds in a Year.

To sum up Harrison's pendulum technology it would seem that he varied, where possible, the many factors which affected the periodicity of the pendulum so that the sum of their disturbing influences would cancel each other out.

Although many after him used his gridiron pendulum to compensate for temperature fluctuation, none took the various other factors into account such as the correct size of the driving weight and the effect that changing barometric pressure or temperature would have on the resistance of the air to the passage of the pendulum, including the frictional resistance and the "flotation factor" (the decrease in the effective weight of the pendulum with increased pressure). An example of the way in which some of these may cancel each other out is that whereas a rise in temperature will, in an uncompensated pendulum, slow it because of its increased length, it will also decrease the air pressure and thus speed the passage of the pendulum through it.

Harrison's Maintaining Power

When Harrison was working on his first sea clock between 1730-35 he realized that the form of maintaining power used up until that time (bolt & shutter), which required setting manually, would not be suitable and thus he devised a new form which would automatically supply power to the train as soon as winding commenced. This became known as "Harrison's Going Ratchet" or just Harrison's maintaining power and was to be adopted by most regulator, as well as other manufacturers although "bolt & shutter" never completely went out of fashion.

Harrison's Unfinished Regulator

Although John Harrison's precision pendulum clocks, already referred to, were designed and made between 1723-1730, he produced at least one (and possibly two) others in later life whilst living in London. This survives and is still with The Royal Astronomical Society (Fig. 10-11) to whom it was presented by Harrison's descendants in 1836. It is unlikely that he finished adjusting it before he died. A new case, pendulum and weight were made for it when Lt. Commander Gould restored it in 1928.

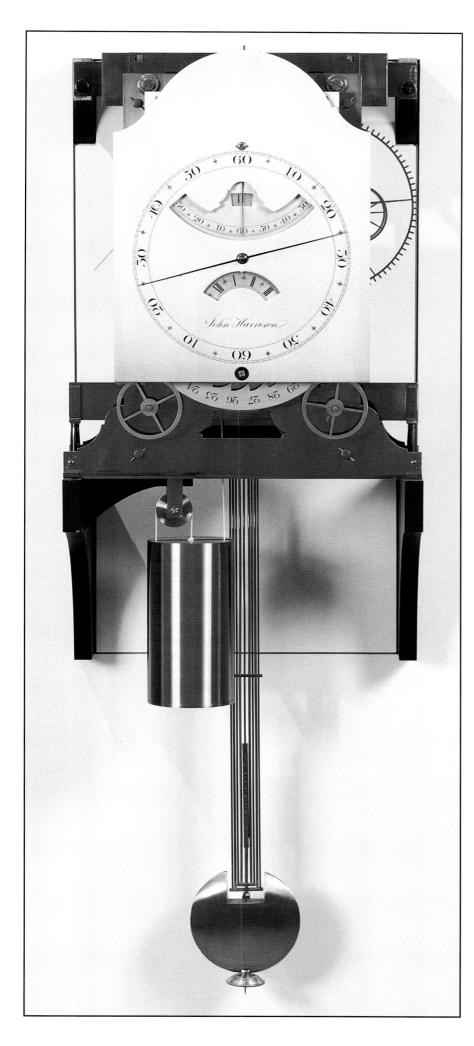

Pages 211-215
Figs. 10-11A, B, C, D, E. **John Harrison's last regulator, which is believed to have been uncompleted at the time of his death in 1776.** At the time it was photographed the pallets and their arbor and the crutch had been removed and the train secured.

This regulator still makes use of Harrison's anti-friction wheels, roller pinions, lignum vitae bushes and grasshopper escapement, albeit in slightly modified form, but the actual construction is entirely different from his earlier clocks. A high count train is employed with beautifully executed wheels having six crossings and delicate rims. The brass plates, basically rectangular, are spread out at the base.

The finely engraved and silvered 11" brass dial has a center sweep double ended minute hand, a large aperture displaying 30 - 30 for the seconds, a small aperture for the hours just below the dial center and a third aperture below 60 for the date.

It is thought that this regulator was unfinished at the time of his death and that the remontoire now fitted was the work of another clockmaker, but by whom and when are not known.

The regulator remained with Harrison's family until 1836 when they presented it to the Royal Astronomical Society who still own it. It presumably then laid in an uncompleted state for some time until Cottingham carried out restoration work on it. In 1928 Lt. Commander Gould had it finished; writing to Agar Baugh, a specialist in compensated pendulums, on behalf of the Royal Astronomical Society, asking him to make a new gridiron pendulum similar to the one on the 1728 clock in the collection of the Worshipful Company of Clockmakers. This gave rise to some queries because of the brass used[17] by Harrison but in the end Baugh estimated £12 for rods tested at the National Physical Laboratory or £6 for plain rods. The latter were chosen. Baugh also supplied the case, weight, pulley, and winding key. The clock is now on display at the National Maritime Museum.

Interestingly in Burgess's article[16] he draws attention to Harrison's 1775 manuscript which suggests that this may be the second regulator made by Harrison in his later years. The remark is as follows: *"My <u>next or second clock</u> will be somewhat better than if it had been finished sooner."* Burgess also points out that part of a regulator with a gridiron pendulum is visible in Harrison's 1767 painting. Could this be the earlier regulator?

Fig. 10-11B

Fig. 10-11C

Fig. 10-11D

Fig. 10-11E

Harrison's remark made in 1775, the year before his death, would seem to lend credence to the above suppositions[16] and suggests that he had by then completed at least one other regulator: *My next or second clock will be somewhat better than if I had finished it sooner.* (John Harrison. "A Description Concerning Such Mechanism as will Afford a nice, or true Mensuration of Time," p. 52, London 1775.)

John Harrison died at Red Lion Square, London on the 24th March 1776 and was buried in Hampstead Churchyard.

A mark of the great respect Harrison was still held in over one hundred years later was that when, in 1879, The Worshipful Company of Clockmakers heard that his tomb had fallen into neglect they decided to completely rebuild it exactly as it was before, including the inscription on the side, despite the fact that he never joined their Company.

References

[1] Quill, H. *John Harrison, The Man Who Found Longitude.* John Baker, London 1960.

[2] King, A. L. *John Harrison, Clockmaker at Barrow, Near Barton-Upon-Humber, Lincolnshire. The Wooden Clocks 1713-30* contained within Andrewes, W. J. H. (Editor) *The Quest for Longitude,* pp. 164-5, Harvard University, Massachusetts 1996.

[3] The lectures of Nicholas Sanderson the blind Lucasian Professor of Mathematics at Cambridge were never published, they included astronomy, the barometers, hydrostatics, the tides, optics, mechanics and the effect of heat and cold. Harrison's copy was listed in *Bibliotheca Chemico-Mathematics, catalogued works in Many Tongues on Exact and Applied Science,* Vol 2, p. 453, published by Henry Sotheron & Co in London in 1927. Unfortunately, this and other important manuscripts by Harrison have now been lost.

[4] Quill, H. *A James Harrison Turret Clock at Brocklesby Park, Lincolnshire.* Horological Journal Vol. 96, No. 1146 pp. 156-9, and No. 1147, pp. 234-6.

[5] King. A.L. *Op. cit. 2.*

[6] Atkins, S.E., & Overall, W.H. *Some account of the Worshipful Company of Clockmakers of the City of London.* Private publication. 1881. pp. 177.

[7] Burgess, M. *The Scandalous Neglect of Harrison's Regulator Science* contained within *The Quest for Longitude,* edited by W.J.H. Andrewes. Harvard University 1993.

[8] Laycock, W.S. *The Lost Science of John "Longitude" Harrison.* Brant Wright Associates 1976.

[9] Bateman, D. & James, K. *Measurements of the Quality Factor of the pendulums of four clocks by John Harrison.* Antiquarian Horology. September 1985. pp. 479 - 489.

[10] Bateman, D.A. *Vibration Theory and Clocks.* Horological Journal. Vol. 12. July 1977 to January 1978 (seven part series).

[11] Bateman, D. & James, K. *Op. cit.*

[12] Hobden, M.K. *The limiting stability of Pendulums.* Antiquarian Horology. June 1986. pp. 132-133.

[13] Bateman, D. *Reply to Hobden. M.K.*[12]. Antiquarian Horology. June 1986. p. 134.

[14] Daniels, G. *The Art of Breguet.* Fig. 270, p. 241. Sotheby Parke Burnett London & New York 1975.

[15] John Harrison's untitled manuscript dated 10th June 1730 at Barrow, Near Barton-upon-Humber, Lincolnshire. Signed by Harrison (Fig. 10-11).

[16] Burgess, M. *Op. cit.* p. 258.

[17] Thomas, A. *James Harold Agar Baugh, Pendulum Maker.* Antiquarian Horology, Sept 2001, p. 259.

Glossary of Terms

When a word is in italics it indicates that it is defined in the glossary.

Accuracy. Measures the conformity of clock time with some standard, usually *Co-ordinated Universal Time*. *Precision* is a pre-requisite for accuracy, but is not in itself sufficient to achieve it.

Amplitude. Scientific term for the semi-arc of swing of a pendulum or balance, i.e. the angle between centre and extreme of swing.

Anchor Escapement. The anchor or recoil escapement has been in common use since it is believed to have been first used by Joseph Knibb, c. 1669. It was initially employed on turret clocks and longcase clocks and from c. 1800 in bracket and wall clocks. The basic principles of the anchor were incorporated by Mudge in the lever escapement which was applied to watches and spring driven travelling clocks, hence the common use on the Continent of the term "anchor" when referring to a lever escapement. It consists of a wheel which is both arrested by and gives impulse to the anchor via the inclined faces of the pallets on either end and it is called a recoil escapement because when the teeth of the escape wheel fall onto the faces of the pallets they are forced back until the pendulum or balance reaches the end of the swing. The wheel then moves forward again and in doing so provides impulse.

Anti-Friction rollers. These were sometimes used in an attempt to reduce friction or prevent undue loading on a pivot by supporting the arbor with rollers. Harrison was one of the first to use them on his famous "sea clocks" and Reid and Denison both employed them on regulators, mainly at the bottom end of the train where the load on the pivots would be heaviest. They were also occasionally used to support the arbor or staff of the balance wheel. In the 18th century they were often referred to as friction rollers; i.e. to reduce friction.

Arbor. This is the axle on which the wheels and pinions are mounted. In the latter case they are often formed as one piece. The arbor for a balance is usually referred to as a staff.

Arc. This is the angle through which a pendulum or balance wheel oscillates or vibrates.

Armature. A piece of soft iron which is attracted by an *electromagnet*.

Astronomical clock or watch. a) One which indicates the movement of some of the heavenly bodies relative to each other or b) One which indicates the Astronomical day, which is 23 hours 56 minutes and 4 seconds long (as measured by *mean solar* or clock time) c) In the USA the term astronomical or "astro" clock is often used to refer to a longcase regulator (precision pendulum clock) when the dial is laid out with a centre sweep minute hand and large seconds and hour rings usually, respectively below 12 o'clock and above 6 o'clock.

Banking. A pin, peg or other means of limiting the extent of motion of any moving part.

Barometric Error. The variations in the timekeeping of a clock caused by rising or falling barometric pressure which alters the density of the air through which the pendulum has to pass and also its buoyancy and thus its speed of oscillation. This is normally only of significance with Precision Pendulum Clocks (Regulators).

Barrel A. Mainspring. The circular container which houses the mainspring. There are three main types: a) "Going" when the barrel is integral with the great wheel which imparts the power to the train of wheels as it rotates. b) "Standing" when the barrel is fixed to one of the plates of the movement and the arbor, which is attached to the centre of the mainspring, rotates. c) "Fusee" when it is connected via a line or chain to a fusee.

Barrel B. The drum onto which the line is wound on weight-driven clocks.

Barrel Arbor. The axle or arbor inside the barrel onto which the mainspring is fixed at its centre.

Bezel. The grooved ring of a watch or clock which holds the glass protecting the dial.

Bi-metallic (compensation). There are two main methods of bi-metallic compensation (a) Two dissimilar metals fixed together, usually by pinning, as in the gridiron, so that their expansion and contraction due to changes in temperature cancel each other out and the effective length of the pendulum stays constant and thus timekeeping is unaffected. (b) Two different metals fixed together to form a bi-metallic strip. Because of their uneven expansion rates the strip, either supporting a pendulum or in a balance, will curve and this movement is used to achieve thermal compensation

Blanc. A partially completed clock movement including plates and spring barrels.

Blanc-Roulant. The basic movement of a clock including wheels and pinions which will require finishing and various other components such as the escapement, dial and hands adding.

Blueing. The process of turning steel blue by heating it to a high temperature and then usually wiping it with oil whilst still hot to protect it. It is used for most steel clock hands and also for other components such as screws and clicks, partly to make them more attractive but also to protect them from corrosion.

Bob. The weight at the bottom of a pendulum, so called because it once bobbed in and out of each side of the case.

Bolt and Shutter. A form of maintaining power to keep the clock going while it is being wound and the driving power is removed.

Bridge. A two ended support for a pivot.

Cannon Pinion. That to which the minute hand is attached and which also drives the motion work.

Centre Seconds. The provision of a long seconds hand sweeping out from the centre of the dial.

Chapter Ring. The ring or dial on which the minutes and hours are marked out.

Chiming clock. One which sounds the quarter hours by playing a tune on a set of four or more bells or gongs.

Chops. Jaws gripping the top of a pendulum's suspension spring.

Chronometer. This term is used to denote a very accurate clock or watch and more specifically one employing a *detent escapement*. There are two main types: "the pocket chronometer" which is a watch employing a *detent escapement*, and the "boxed ship's chronometer" in which the movement is suspended by gimbals in a three tier box to allow for the motion of the ship. The chronometer escapement is seen quite frequently on fine English carriage clocks such as those made by Dent and Frodsham.

Circular error. In scientific terms, this is a losing rate caused by the free vibration of a pendulum in a circular arc through a non-zero angle. The losing rate is 1.645 seconds per day when the extremes of swing are one degree each side of center, a loss which increases as the square of the angle. If the angle of swing stays constant, the error will be annihilated by regulation.

Clepsydra. A clock which measures the passage of time by means of the flow of water.

Click. In electrical horology, a pawl attached to the pendulum which advances the *count wheel* (q.v.).

Cock. A single ended bracket carrying a pivot.

Collet. A circular ring which holds a component in place, for instance a wheel on its arbor or the minute hand on to its pipe, when it is sometimes referred to as a hand collet.

Comtoise or Morbier Clock. A wall or less commonly a longcase clock made in the Franche-Comté and characterised by striking on the hour and again at two minutes past, a feature it shares with many public clocks in France, particularly those made in the East.

Constant Force Escapement. One in which a constant or even force is applied to the pendulum or balance at each oscillation by a weight which is always lifted or a spring deflected by exactly the same amount prior to being released.

Contrate Wheel. This has its teeth set at right angles to the plane of rotation of the wheel and is used to transfer the drive through 90° as, for instance, when connecting the going train of a carriage clock to the lever or cylinder escapement.

Co-ordinated Universal Time (UTC). Has the same rate as *International Atomic Time* (TAI), but is adjusted by leap seconds about once a year so that the difference from *Universal Time* (UT1) is never more than 0.9 seconds. *International Atomic Time* is based solely on atomic clocks, while Universal Time is based on the Earth's rotation. Since 1972 all radio stations have transmitted UTC.

Count wheel. A. A method of controlling the strike used on all early clocks and eight day longcase and bracket clocks up until c. 1700 and is also sometimes seen on carriage clocks prior to 1850. It employs a disc with notches at varying distances from each other at its periphery. The greater the distance the higher the number of hours struck. Unlike rack striking it is not synchronised with the time train and thus cannot be used with repeat work. A different use of the term count wheel applies to the ratchet wheel employed for counting the swings of a pendulum between impulses, required only for use with escapements which do not operate at every swing. **B.** In electrical horology, a ratchet wheel which is advanced, one tooth at a time, by the pendulum. An arm (or arms) on this wheel triggers the impulse to the pendulum by releasing a gravity arm or operating a switch.

Crossings. The spokes of a wheel.

Crown wheel. The escape wheel in a *verge escapement*.

Crutch. This connects the pallet arbor, to which it is fixed, to the pendulum, the top of which usually passes through a loop or fork in it.

Cycloidal. Following the mathematical curve known as a cycloid.

Dead Beat Escapement. One in which the escape wheel does not recoil. Generally used on regulators and other *precision clocks*. It was probably first employed by Tompion and Townley, for the year clocks installed at Greenwich in 1676 but the final version, which was universally adopted, is credited to George Graham c. 1715.

Detached Escapement. One in which the pendulum or balance is free from the influence of the escape wheel and thus the movement, except during locking and unlocking.

Detent. This is a locking device which generally acts by engaging one of the teeth of a wheel, as in maintaining power. It is applied in particular to the locking piece of the *Detent escapement*.

Detent Escapement. A precision escapement used in chronometers in which impulse is only given to the balance at every other oscillation. Its principle advantage is that the balance swings almost entirely free of the influence of the escape wheel except when it makes contact with the passing spring or receives impulse.

Dutch Striking. When the clock strikes the hour in full at the hour and repeats it at the half hour, using a high toned bell at the halves and a lower toned one at the hour.

Electromagnet. A *solenoid* (q.v.) with a soft-iron core.

Enamel. A vitreous material fired onto various metals such as gold, silver, brass and iron.

Endplate (or stone). A cover made of steel and sometimes set with a jewel stone, designed to limit the end play of an arbor and reduce friction by preventing its shoulders butting onto the plates.

Engine Turning. A system of decorative repetitive machine engraving.

Epicyclic Gearing. The word epicycle is derived from the Greek and means "upon a circle". Epicyclic or 'Sun and Planet' gearing is the use of a sun or central wheel with planetary wheels which run around it and an outer annular ring with teeth on its inner aspect.

Equation of Time. The difference between *solar* and *mean (solar)* time which may be as much as 16 minutes. Four times a year the two coincide.

Equation Tables. These tables, which show the variations in the *Equation of Time* throughout the year, are or were used to check the clock's time by using a sundial to find *solar time* and then adding the Equation of Time onto or subtracting it from this.

Error. Is the difference in seconds between clock time and some standard time. The standard used may be Co-ordinated Standard Time, the length of the sidereal day, a stand-alone signal generator, or even another clock.

Escapement. The mechanism which controls the speed of rotation of the wheels in the time side of a clock and thus the hands connected to them. It consists of a balance or pendulum which every time it completes a swing (or sometimes two swings) releases one tooth of a wheel known as an escape wheel and thus allows all the rest of the wheels to rotate a set amount. At the same time the tooth imparts an impulse to the pendulum or balance, which keeps it in motion.

Fire-Gilding. This is the gilding of material such as silver or brass by applying an amalgam of gold and mercury to it and then driving off the mercury by heat.

Flicker floor. A term borrowed from electronics and denoting the optimum level of timekeeping stability for any given clock, assessed by long-term testing and analysis.

Flirt. This is used to provide sudden movement, for instance to calendarwork, so as to make it change at exactly the right time. It is also used in repeating mechanisms to ensure that the rack tail drops at the right time onto the correct position on the *snail* and is employed for the operation of the strike in many 18th and 19th century English bracket clocks.

Fly. A fan, usually two or three bladed, which is used to regulate the speed at which a clock strikes or chimes. It is also occasionally employed in other circumstances, such as the rewinding of a *remontoire* and in *gravity escapements*.

Foliot. A pivoted bar with adjustable weights, used on early clocks with verge escapements instead of a balance.

Frame. The plates and pillars of a watch or clock movement.

Free Pendulum. One which is detached from any disturbing influences of the movement.

Frequency. The number of *periods* of vibration per second (0.5 hertz for a seconds pendulum). Frequency in hertz is the reciprocal of period in seconds.

Fuzee (or Fusee). A spirally grooved cone used to equalise the decreasing force of the mainspring as it runs down. Its profile is varied to match the characteristics of the mainspring.

Gathering Pallet. This is used in rack striking, a single leafed pinion "gathering up" one tooth of the rack each time the clock strikes.

Gimbals. A pivoted ring into which an item such as a ships' chronometer is pivoted so that no matter what angle the ship takes up the chronometer will remain horizontal.

Grande Sonnerie. At each quarter the hours are struck first followed by the quarter and at the hour just the hour is usually sounded.

Gravity Escapement. An escapement in which a constant force is imparted to the pendulum by lifting up a weight a set distance and then releasing it onto the pendulum to give it an impulse. The most famous example is the double three-legged gravity escapement designed and used by Lord Grimthorpe in the Westminster Palace clock known as Big Ben.

Great Wheel. The first and largest wheel in a clock or watch train.

Greenwich Mean Time (GMT). Is essentially the same as *Universal Time* but has now been replaced for practical purposes by *Co-ordinated Universal Time*.

Gridiron pendulums. Have a rod built of separate bars of two metals with different coefficients of expansion, so that the length of the pendulum remains the same even though the temperature changes. John Harrison used a total of nine bars of brass and steel when he first introduced the design in 1729. Some later makers have used fewer rods and other metals.

Guilloché. The decoration of an item with *engine turning* which may be overlaid with translucent *enamel*.

Hardening. The heating and then rapid cooling of steel to produce a hard surface resistant to wear. Different techniques may be used to harden various other metals.

Helical Spring. A spring with its coils one above the other in the form of a cylinder.

Hour Wheel. That to which the hour hand is attached.

Impulse. The force applied at regular intervals to a balance or pendulum to keep it swinging.

International Atomic Time (TAI). Is based on a worldwide group of atomic clocks set to agree with *Universal Time* (UT1) on 1st January 1958.

Invar. An alloy of nickel and iron invented by Dr. Guillaume in 1897 which has an extremely low coefficient of expansion or contraction. It is used in pendulums and balances in various forms to minimise any effect on timekeeping due to changes in temperature.

Isochronous. This means moving in equal time, for instance in relation to the beat of a pendulum.

Jewelling. Jewels are sometimes used in precision clocks and more generally in watches because of their extreme hardness and ability to take a high polish. Their primary purposes are to reduce friction and wear. Thus they are used for the acting faces of pallets, pivot holes and as endstones to control the end float of the arbors and stop their shoulders butting on the plates.

Jumper. A spring which is shaped so that it makes a piece move suddenly, for instance the *star wheel* carrying the hour snail such as is found on a repeating French carriage clock. This enables the clock to always repeat the previous hour correctly right up until the last minute before the next hour is struck. See also star wheel.

Jumping hours and minutes. A clock or watch in which the hour hand, for instance, jumps from hour to hour rather than moving continuously.

Lacquer. A clear covering used to protect metals and prevent tarnishing. It also refers to the painted decoration, originating in the Far East, which was applied to furniture and other items, including clock cases.

Lantern Pinion. One which is formed by wire pins held between circular plates.

Leaf. The tooth on a *pinion*.

Lever Escapement. The escapement most generally used on watches and portable clocks. It was invented by Thomas Mudge c. 1757 and employs a lever to inter-connect the pallets and balance.

Locking Plate. See countwheel.

Mainspring. The principal spring of a watch or clock which provides the driving force.

Maintaining Power. During winding the power is removed from the train through which it is transmitted to the escapement and thus the direction of rotation of the wheels is reversed unless going barrels or Huygens endless chain have been employed. To prevent this and keep the clock going "maintaining power" is sometimes provided in which an auxiliary spring or weight, is brought into operation to impulse the escapement during winding.

Manometer. A mercury-filled glass U-tube, open at one end, and used to measure the pressure of a gas, like a barometer.

Mean Time. More correctly known as "Mean Solar Time" it is the average of all the *solar days,* which are constantly varying in length throughout the year.

Mercury compensated pendulums. Have a steel rod and a mercury-filled cylindrical iron or glass bob. The length of the mercury column increases with temperature, counteracting the expansion of the steel rod and keeping the effective length of the pendulum constant. It was invented by George Graham in 1726.

Millibar. A scientific measure of barometric pressure. One bar is almost exactly 750 mm of mercury.

Motion Work. The wheels, usually situated immediately behind the dial on the front plate of the movement, which gear down 12:1 so that the hour hand moves at the correct speed relative to the minute hand.

Movement. All the mechanical components of a clock or watch.

Multiple regression analysis. Is a technique to obtain the linear equation which best fits a set of data containing more than one independent variable.

Ormolu. Firegilt bronze decoration.

Pallets. The part or parts of the escapement upon which the teeth of the escape wheel act to provide both locking and impulse.

Pendule Sympathique. A combination of watch and clock devised by Breguet in which the clock automatically sets the watch to time and also winds or regulates it.

Period. The time for a complete cycle of vibration (two seconds for a seconds pendulum). Period in seconds is the reciprocal of *frequency* in hertz.

Perpetual Calendar. When a correction is automatically provided for the variation in the number of days in the month and year.

Photoemissive cell. A device in which light produces a current by emitting electrons.

Pillars. The distance pieces holding the two plates of the movement rigidly together.

Pinion. A small toothed wheel in a clock or watch, often made integral with the arbor which carries it. Its teeth are usually called leaves.

Pin Pallet (escapement). In this design the pallets are in the form of pins.

Pin Wheel. An escapement in which the teeth of the escape wheel are replaced by pins projecting out of the wheel at right angles. There are many different forms, with pins sometimes projecting from both sides of the wheel.

Pivot. The reduced and finely polished end of an arbor which rotates in a hole in the plates of the movement.

Pivoted detent. A chronometer escapement in which the *detent* is pivoted and controlled by a spring.

Precision. Is a measure of the stability of clock *rates*. A clock which gains *exactly* a minute a day, every day, is extremely precise. But it cannot be described as accurate, since it will never tell the correct time.

Precision clock. A clock intended by its maker to reach the highest standard of precision and be capable of accurate timekeeping.

Quartz clock. A clock in which the timekeeping is controlled by the vibration of a quartz crystal.

Rack Striking. This was invented c. 1675 and unlike the countwheel enables the striking to stay correct relative to the time. This permitted repeat work to be provided on clocks. The mechanism consists of a segment with teeth on it which is dropped back just before the hour an amount which varies with the number of hours to be struck. At the hour a rotating tag known as a gathering pallet advances the teeth one at a time, each one corresponding with one strike.

Rate. The amount by which a clock gains or loses in a given period of time, usually a day.

Rating Nut. The nut immediately below the pendulum bob which is rotated to raise or lower it and thus regulate the timekeeping of the clock.

Regulator. This is used both to describe a precision pendulum clock and is also an alternative name for an index used to regulate balance wheel clocks by varying the effective length of the hairspring.

Remontoire. A device which winds and releases a small weight or auxiliary spring at regular intervals, usually between 30 seconds and five minutes. This provides either a) A constant force on the escapement and thus the pendulum or balance, thereby improving timekeeping, b) Only one mainspring may be provided and part of the power taken from this via the remontoire to drive a second train. This may be for the strike or time.

Repeat Work. The mechanism by which a clock may be made to strike the previous hour or quarter, for instance when a button is depressed or a cord pulled.

Repoussé. Decoration in relief, usually in brass, silver or gold, which is created either by means of hammering out the design with punches or using dies and counter dies.

Rococo. Elaborate curved and scrolling decoration, originally inspired by sea shells and rock formations.

Short circuit. A path of low resistance.

Sidereal Time. The sidereal day is the elapsed time between two successive passages of the same star across a local meridian. It is 3 minutes 56 seconds shorter than the mean *solar day*.

Set Up. The power which is still present in a mainspring even after it has run down and the clock has stopped, provided stopwork is fitted.

Silvering. The chemical deposition and polishing of silver on components such as the *chapter ring* of a dial. This should not be confused with electroplating.

Snail. A stepped cam which regulates the number of hammer blows struck in *rack striking*.

Solar Day. This is the time recorded between two successive passages of the sun over the meridian. The length of the solar day varies throughout the year by up to 30 minutes because of the Earth's elliptical orbit round the sun, and its tilted axis of rotation..

Solar Time. That indicated by the sun. It varies by up to 16 minutes either side of *mean time* throughout the year.

Solenoid. A cylindrical coil which acts as a magnet when an electric current passes through it.

Spotting. A speckled finish in the form of a succession of circles which is most commonly seen on the backplates of English chronometers and sometimes mantel clocks from c. 1850 onwards.

Star Wheel. A star shaped wheel which is located by a spring with a vee shaped end known as a jumper. Once the tip of a tooth on the wheel has passed the tip of the spring it will automatically be pushed forward until the jumper rests in the space before the next tooth.

State of Wind Dial. This indicates the amount by which the mainspring is wound. It is also known as an "up and down" indicator.

Stop Work. This limits the amount by which a mainspring or weight may be wound or allowed to run down, thus avoiding the use of the spring when it would be too weak or too strong, which would affect the timekeeping. It is also quite often used on weight driven regulators.

Suspension Spring. The strip of steel used to support the pendulum and allow it to swing.

Thermionic valve. A vacuum tube which can be used to amplify electric currents. Because of its high power consumption it was rarely used in electric clocks.

Timepiece. A clock which does not strike.

Ting-Tang striking. The sounding out of the quarter hours on two gongs or bells, one of a higher note than the other.

Tourbillon. A revolving cage which carries the escapement. Invented by Breguet in the late 18th century to even out the errors caused by the effects of gravity upon the balance.

Train. A series of intermeshing wheels and pinions. On a timepiece, for instance, they connect the spring barrel with the escapement.

Transistor. A solid state device which can be used to amplify an electric current. Because its power requirements are modest it is widely used in electric clocks.

Universal Time (UT). Is based on the rotation of the Earth. It is defined as *mean solar time* at Greenwich + 12 hours. In practice the mean solar time is calculated from Greenwich Mean *Sidereal Time*, and called UT1.

Verge escapement. The earliest escapement used on mechanical clocks. The axis of the pallets is at right angles to that of the escape wheel.

General Index

Accuracy, 163, 217
Accuracy of Harrison's regulators, 207
Airy, G.B. 12, 57, 128, 131, **4-7**
Aluminum, 74
Amplitude, 217
Aneroid compensation, 103
Anti-friction rollers, 200, 204, 211, 217, **10-5E**
Anti-friction wheels, 60
Apparent time, **9-7A,B**
Apparent solar time, **9-6**
Arbor, 217
Arc, 217
Armature, 217
Astrolabe, 174. **9-5A,B,C**
Astronomical clock or watch, 217
Astronomical observatories, 163
Astronomy, 9
ATO clock, 145
Banking, 217
Barograph. Clockwork, 122
Barometric compensation, 52 - 54, 101, **4-2, 4-3, 4-4**
Barometric error, 217
Barometric pressure, 52, 53, 54, 77, 210, **8-1**
Barrel, 217, 218
Barrel Arbor, 218
Belmont, 16
Bentley Regulator, 141
Bertele, Hans von, 14
Bezel, 218
Big Ben, 46, 131, **3-18**
Bimetallic bars, **5-31**
Blacksmiths, 169
Blanc, 218
Blanc-Roulant, 218
Bliss, Nathaniel, 10
Blueing, 218
Bob, 218
Bothkamp Observatory, 165
Boxwood, 194
Bradley, James, 10, 11
Brahe, Tycho, 9
Bridge, 218
Brocklesbury Park, 115, 194, 207, **10-4, 10-5A-K**
Calendar:
    Gregorian, 27, **3-12**
    Julian, 27
Cannon pinion, 218
Carbon fiber, 105
Centre seconds, 218
Ceramic quartz, 103
Chapter ring: 218
    twenty-four hour, **3-12**
Chiming clock, 218
Chops, 218
Chronometer, 218
Circular deviation, 210
Circular error, 51, 218

Cleopatra's Needle, 14
Clepsydra, 14, 218
Click, 219
Clock stars, 11
Cock, 219
Coefficient of expansion of wood, **5-34**
Collet, 219
Comptoise or Morbier clock, 219
Contrate wheel, 219
Coordinated universal time (UTC), 163, 219
Count wheel, 219
Crepuscular line, 175
Crossings, 219
Crown wheel, 14, 219
Crutch, 219
Cycloidal, 219
Cycloidal cheeks, 15, 51, 112, 116, **4-1**
Cycloidal curve, 15, 51
Declination, 9
Definition of Time Act, 50
Degree clock, 23
Detent, 219
Dolcoath Copper Mine, 57
Double dialled clock, **3-7**
Drayton Tompion, **9-6**
Dutch striking, 220
Earth battery, 141
Earth, Density of, 57
Electrical Clock. Alexander Bain, **7-1**
Electric telegraph, 12, 44
Electricity, 12
Electrically rewound movements, 140
Electromagnet, 220
Electronic switching, 145
Elinvar, 100
Emerson's Table, 55
Enamel, 220
Endplate (or stone), 220
Energy loss from the pendulum, 166
Engine turning, 220
Epicyclic gearing, 220
Equal time, **9-7A,B**
Equation clocks, 174, **9-7A,B**
Equation disc, **3-13A,B,C**
Equation manual, **3-9A,B,C**
Equation tables, 14, 18, 22, 27, **3-1, 3-3**
Equation table, Harrison's, 204
Equation of time, 220, **9-7A,B**
Equation of Natural Days, 18, 22, 24, **3-5A,B**
Error, 220
Escapement remontoire, 60
Escapement, 220
Escapements: 111
    Amant's pinwheel, **6-6A**

Anchor, 15, 16, 217, **6-3**
Balance wheel, 111
Berthoud's detached, **6-14A**
Bloxham's, 130, **6-17B**
Breguet's detached, **6-14B**
Centrifugal governor, **6-21A**
Constant Force, 120, 126, 219
Coup Perdu, 118, **6-6D**
Cross beat, 9, 14, 112
Cumming's gravity, **6-8**
Dead beat, 16, 17, 114, 219, **6-4**
Dennison's experimental detached, **6-18A**
Dennisons four legged gravity, **6-19**
Detached escapement, 219
Detached gravity, 122
Detent, 219
Duplex, **6-16**
Foliot, 111
Free pendulum, 132, 136
Graham Dead beat, **6-4**
Grasshopper, 200, 203, 204, 211, **10-5E**
Gravity escapement, 120, 132, 221
Hardy's spring pallet, **6-13**
Harrison's "Grasshopper," 115, 116, **6-5A, 6-5B**
Horizontal, 180
Kelvin's free pendulum Seconds', **6-21C**
Kelvin's half seconds', **6-21B**
Lepaute's pinwheel, **6-6C**
Leroy's, **6-24**
Lever, 222
Massey's, Spring pallet, **6-11**
Mudge's gravity, **6-9**
Nicholson's, **6-10**
Nicholson's gravity, 105
Pinwheel, 118, 119, **6-6D**
Reid's spring pallet, **6-12**
Remontoire, 120, 128, **6-15A**
Airy's, **6-15B**
Train, 120
Riefler, **6-22A**
Strasser's, **6-23A, 6-23B**
Thiout's separate pallet, **6-7A**
Verge, 224
Verge and foliot, 111
Expansion of different metals, 52, 82
Fire clocks, 14
Fire-Gilding, 220
Fitzwilliam Museum, Cambridge, 175
Fixed stars, 210
Flamsteed, John, 10, 17, 20 - 23, 27, 114, 169, **2-3**
Flicker floor, 220
Flirt, 220

Flotation factor, 54, 210
Fly, 220
Fly back, **3-11A,B**
Foliot, 14, 220
Frame, 220
Frequency, 220
Friction, 60
Fused silica, 103
Fuzee (or Fusee), 221
Galileo, 9, 15, 51
Gathering pallet, 221
Gershom Parkington Museum, 16
Gimbals, 221
Glass, 103
Glasses, Ultra low expansion, 105
Good, Richard, 21
Grande sonnerie, 221
Gravitational Attraction of the Sun & Moon, 58
Gravity, 55, 58
Gravity arm, 155, 156
Gravity impulse, 133, **6-20**
Gray, Lord, 124
Great wheel, 221
Greenwich Mean Solar Standard clock, 149
Greenwich Mean Time (GMT), 221
Greenwich Time Signal, 149
Guilloché, 221
Hall, Professor, 60
Hardening, 221
Harrison's 1730 Manuscript, 209, 210
Harrison's Unfinished regulator, 210
Harton Colliery, **4-7**
Helical spring, 221
Hevelius, 112
High Water, 175
Hipp observatory clock, **7-4A, 7-4B**
Hipp toggle, 141
Hit and miss Synchronizer, 156
Hooke, Dr., 15, 16, 114, 169
Horological Disquisitions, Smith, John, 15
Horrox papers, 27
Hour wheel, 221
Humidity, 52, 58, 98
Huygens, Christiaan, 9, 15, 27, 51
Huygen's endless rope, 62, **4-11**
Impulse, 221
International Atomic Time (TAI) 221
International Time Standard, 50
Invar, 221
Isochronal, 18
Isochronicity, 51
Isochronism, 15
Isochronous, 15, 221
Jewelling, 60, 221
Journeyman or "Assistant" clock, 10

Jumper, 221
Jumping hours and minutes, 221
Jupiter's satellites, 22
Kater, 56
Kinfauns Castle, 124
King's College, Cambridge, 16
Knife edge, 201
Knife edge suspension, 58, 200
Lacquer, 222
Lantern pinion, 222
Laycock, 76
Leaf, 222
Leclanché cell, 144
Leicester, Earl of, **2-4A,B**
Lignum vitae bushes, 200, 201, 211, **10-5B**
Liverpool, 49
Locke, John, 12
Locking plate, 222
Longitude, 9
Lowther, Sir James, **2-4C,D**
Lubrication, 60
Lunar-distance, 9
Magnets:
    Oscillating, 152
    Rotating, 152
Mainspring, 222
Maintaining power, 61, 222
    Bolt & shutter, 62, 218, **4-12A,B**
    Epicyclic, 65, **4-14A,B**
    Harrison's, 64, 210, **4-13A,B,C**
    Turret clock, **4-10**
Mannheim Observatory, 164
Manometer, 222
Maskelyne, Nevil, 10, 11
Mean time, 222
Mean time clocks, **9-8A-D**
Mean and solar time clocks, **9-15A,B,C**
Mercurial compensation, 67
Millibar, 222
Moons of Jupiter, 9
Moore, Sir Jonas, 18, 22, 23, 169, **2-4A-D, 2-5**
Morbier clocks, 15
Motion work, 222
Movement, 222
Multiple regression analysis, 222
NASA, 12
National Physical Laboratory, 165
Naval Observatory, Washington, 165
Navigation, 12
Neuchâtel Observatory, 139, 142, 165
Newton, Sir Isaac, 188
Nostell Priory, 191
Octagon Room, 10, 21, **2-4A-D, 2-7**
Old Royal Observatory, 23
Ormolu, 222
Orrery, Lord, 180
Pallets, 222
Pelham, Sir Charles, 194
Pendule sympathique, 222

Pendulums:
    Compensated, 88, 180, **5-14**
        Bimetallic, 218
        Buckney's, 91
        Ellicott's, 83, **5-19**
        Hall's, **5-27**
        Reid's, **5-25A,B**
        Ritchie's, 96, **5-31, 5-32**
        Troughton's mercurial, **5-7**
        Ward's compensated pendulum, **5-24**
        Zinc & steel, 90
    Electromagnetic, 140, 141
    Ferro-nickel, 99
    Free, 60, 133, 220
    Gridiron, 52, 75, 76, 77, 79, 81, 203, 204, 208, 221, **5-8, 5-11, 5-12, 5-15A,B, 5-16, 5-17A,B**
        Harrison's gridiron, **10-8**
    Invar, 61, 99, 103
        Fix-invar, 102
        Super-invar, 102, 103
    Kater's, 57, **4-6**
    Master, 133
    Mercurial, 52, 72,
    Mercury compensated, 141, 222, **5-1**
    Nickel-Steel, 103
    Quartz, 103, 104
        Frederic Ecaubert, **5-40A**
        Reubold's ¾ seconds, **5-42**
        Satori's, **5-41A,B**
        Quartz rod, 105
    Rhomboidal, 93, **5-30A,B,C**
        Hooke's rhomboidal, 93
        Troughton's rhomboidal, **5-29**
    "Royal," 16
    Slave, 133, 156
    Wood rod, 98, 99, **5-33, 5-35**
    Zinc & steel, 91, **5-28A,B**
Pendulum's support, 59
Period, 222
Perpetual calendar, 222
Phipps, Captain, 55
Photocell, **7-8**
Photoemissive cell, 145, 222
Pillars, 223
Pinion, 223
Pin pallet, 223
Pin wheel, 223
Pivot, 223
Pivoted detent, 223
Planetarium, 180
Point suspension, 59
Pond, John, 10
Portable regulator, **4-8**
Positional astronomy, 9
Potsdam, 166
Precision, 163, 223, **8-2**
Precision clock, 223
Quality factor, 207
Quartz, 103,
Quartz clocks, 146, 161, 223

Rack striking, 223
Railway timetables, 44
Rate, 223
Rating a clock, 163
Rating nut, 223
Regulation of Precision clocks, 107
Regulating nuts, 107
    Riefler's regulating nut, **5-44**
Regulation by Barometric Pressure, 109
Regulator, 223
Remontoires, 60, 223 **3-10A,B**
    Electric, 140
    Gravity, 149
    Spring, 148
Repeat work, 223
Repoussé, 223
Richter, Jean, 55
Riefler's barometric compensation unit, **5-38**
Riefler's compensation, 100, **5-36**
Riefler's pendulum, **5-37**
Rock crystal clock, 14
Rococo, 223
Roller pinions, 200, 204, 207, 211, **10-11A-E**
Royal Astronomical Society, 210, **10-11A-E**
Royal Mail, 44
Royal Observatory Greenwich, 17, 44, 47, 164, 169, **2-2**
Royal Society, 17
Saunderson's lectures, 191
Selenium plate, 145
Set up, 224
SH.O, 156
Shadow clocks, 14
Shepherd's Galvano-Magnetic Clock System, 47
Short circuit, 223
Shortt's clock, 156
Shortt's Duplex clock, 155
Shortt-Synchronome free pendulum clock, 155, 158, **7-20B, 7-20C**
Sidereal Tables, **3-21A,B,C**
Sidereal time, 223
Sidereal time clocks, **9-8A-D, 9-13A,B**
Signs of the zodiac, **3-8A,B, 3-13A,B,C**
Silvering, 224
Slave clock, 156
Smith, John Horological Disquisitions, 15
Snail, 224
Solar Day, 24, 224
Solar time, 224,
Solar Time Clocks, 27
Solenoid, 224
South Shields Coal Mine, 57
Soviet Union, 160
Spiral balance spring, 169
Spotting, 224
Stage coaches, 44
Standard deviation, 163, 164
Star transits, 164

Star wheel, 224
State of wind dial, 224
Steam locomotive, 44
Stop work, 224
Stratified temperature compensation, **5-38**
Stratification, 61, 102
    Compensation for, 103
Stratified temperature, 52, **5-39**
Sundials, 14, **3-2A,B**
Sunrise/sunset shutters, **3-12**
Suspension spring, 58, 224
Synchronome switch, 155
Table regulators, French, 79, **5-15A,B**
Tank clock, **5-48**
Temperature, 210
Thermionic valve, 224
Time:
    Apparent Solar, 10
    Equal, 24
    Equation of, 10, 17, 18, 24, **3-3, 3-4, 3-6, 3-8A,B, 3-10A,B**
    Greenwich Mean, 24, 44, 49, **3-20, 3-21A,B,C**
    Liverpool Mean, **3-21A,B,C**
    Local, 24, 44, 49
    Local mean, 47, **3-20**
    Local mean solar, 44
    Local solar, 44
    Mean, **3-15A-E, 3-16A-D**
    Mean Solar, 10, 11, 24
    Sidereal 24, 40, **2-8, 3-15A-E, 3-16A-D**
    Solar, 24, 27, 37
Time balls, 44, **9-13A,B**
Time Ball Tower, Deal, **3-17A,B,C, 7-14**
Time stars, 40, 210
Tin whistle, 76, **5-9A,B**
Timepiece, 224
Ting-tang striking, 224
Tourbillon, 224
Train, 224
Transistor, 224
Transit Clock, 10
Transit Instrument, 9
Universal Time, (UT), 224
Wadham College, 16
Water Lane, 169
Wear, 60
Werner, Johann, 9
William III, **3-8A,B, 9-7A,B**
Willis, John, 27
Wren, Christopher, 16, 17, **2-2**
Yarborough, Earl of, **10-4**
Year calendar, **3-4**
Year duration, **9-6, 9-7A,B**
Zulichem, Mr., 15

Makers Index

Further references are made to many of these makers in *English Precision Pendulum Clocks* and *Precision Pendulum Clocks: France, Germany, America, and Recent Advancements* with chapters, either in whole or in part, devoted to those thought to be most important.

Amant, 117
Arnold, John, 77, 164, **5-11, 5-13** (see also *English Precision Pendulum Clocks*)
John Arnold & Son, **5-21** (see also *English Precision Pendulum Clocks*)
Aske, Henry, 180
Bailly, B, Paris.**3-10A,B**
Bain, Alexander, 140, 141, 144, 148, **7-2**
Banger, Edward, 180, **9-8A-D**
Bannister of Colchester, 103 (see also *English Precision Pendulum Clocks*)
Barkley, Samuel, 187, 188
Baugh, Agar, 100
Baumann, **7-11**
Bentley, 141
Berthoud, 60, 61, 81, 99, **5-16A,B, 5-17A,B, 5-18** (see also *Precision Pendulum Clocks: France, Germany, America, and Recent Advancements*)
Bethune, Chevalier de, 119
Bloxham, 130, 131, 132, **6-17A,B**
Bond, 139
Bowell, G. B., 153
Bradley, James, 188
Breguet, 40, 210 (see also *Precision Pendulum Clocks: France, Germany, America, and Recent Advancements*)
Brillié Brothers, 144, **7-5, 7-6**
Brock, James, 131 (see also *Precision Pendulum Clocks: France, Germany, America, and Recent Advancements*)
Brockbanks, 40, **3-15A-E**
Buckney, Thomas, 90, 91
Burgi, Jost, 9, 14, 60, 112, **6-1**
Campani, Matteo, 54
Chamberlain, 137
Clement, William, 15, **6-2A**
Cockey, Edward, Warminster **3-12**
Colley, Thomas, 187, 188
Cooke, London and York, 40, 65 (see also *English Precision Pendulum Clocks*)
Cornu, 144
Coster, Salomon, 9, 15
Cumming, Alexander, 85, 120, 122, **5-21** (see also *English Precision Pendulum Clocks*)
Delander, Daniel, 128, **3-5A,B, 6-16**
Denison, 131, 132
Dent, 58, 59, 90, 91, 128, 165, **4-2, 4-9, 5-3A,B, 6-15A** (see also *English Precision Pendulum Clocks*)
Derham, William, 15, 54, 169, 188
Ditisheim, Paul, 100
Dutton, 187
Ecaubert, Frederic, of New York, 103, 104
Ellicott, John, 52, 83, **5-19, 5-20, 5-21, 5-22, 5-23** (see also *English Precision Pendulum Clocks*)
Etalon, 160, **7-21**
Favarger, 148
Fedchenko, 58, 145, 165 (see also *Precision Pendulum Clocks: France, Germany, America, and Recent Advancements*)
Felton of Elmshorn, **5-10**
General Ferrié, 145, **7-8**
Féry, Professor, 144
Frodsham, 58, 71, 74, **5-4A,B, 5-5** (see also *English Precision Pendulum Clocks*)
Froment, 149, **7-10**
Fromanteel, Ahasuerus, 15
Fromanteel, John, 15
Fromanteels, The, 15
Garnier, Paul, **6-14C**
Graham, George, 10, 11, 17, 52, 55, 67, 68, 108, 114, 116, 117, 164, 165, 168, 180 - 188, **9-9, 9-10, 9-11A,B,C, 9-12. 9-13A,B, 9-14, 9-15A,B,C**
Guillaume, Doctor, 75, 99, 100
Hall, 91, **5-27**
Hardy, William, 68, 125, 165 (see also *English Precision Pendulum Clocks*)
Harrison, James, 115
Harrison, John, 10, 40, 52. 54, 75, 76, 77, 115, 116, 190 - 216, **5-8, 5-9A,B, 10-1, 10-2A,B,C, 10-3A,B,C, 10-4, 10-5A-K, 10-7A-G, 10-9, 10-10, 10-11A-E**
Hatot, Leon, 145, **7-7**
Herz, Adolf, 104
Hirsch, Adolph, 142
Hipp, Matthäus, 141, 142, 144, 165, 166, **7-3**
Hooke, Robert, 93, 112
Hope-Jones, 155, 156
Huygens, 65, 111, 112
Janvier, Antide, 58, **4-8** (see also *Precision Pendulum Clocks: France, Germany, America, and Recent Advancements*)
Johnson, W, Liverpool, **3-21A,B,C**
Joly, August, 149
Kelvin, Lord, 133, 134, 135, **6-21A**
Kelvin & White, 133
Kendal & Dent, 47, **3-19**
Knibb, Joseph, 16, 112
Knoblich, 165, 166
Lavet, Maurice, 144, 145
Lemoine, Paris, 141
Lepaute, 117, 118, **5-23** (see also *Precision Pendulum Clocks: France, Germany, America, and Recent Advancements*)
Le Roy, 54 (see also *Precision Pendulum Clocks: France, Germany, America, and Recent Advancements*)
Leroy, 138, 139 (see also *Precision Pendulum Clocks: France, Germany, America, and Recent Advancements*)
Margetts, 40, **3-16A-D** (see also *English Precision Pendulum Clocks*)
Marrison, 161
Massey, Edward, 123
Matthys, Robert, 108, **5-46, 5-47**
Meath, Earl of, 153
Mercer, Thomas, 153, **7-17B**
Möhren, Paris, **5-22**
Morrison, 165
Mudge, Thomas, 122, 187, 188 (see also *English Precision Pendulum Clocks*)
Nicholson, 97, 122, **5-43A,B,C**
Parr, B, Grantham, **3-11A,B**
Princeps, 149
Quare, 27
Quare & Horseman, 27, **3-3, 3-7**
Reid & Auld, 109, 124 (see also *English Precision Pendulum Clocks*)
Reid, Thomas, 79, 88, 124, **5-25A,B** (see also *English Precision Pendulum Clocks*)
Rentzsch, Sigismund, 68, **5-2AB,C**
Reubold, 105
Riefler, 54, 74, 75, 100, 102, 103, 105, 107, 108, 136, 137, 155, 165, 166, **5-48, 7-9A, 8-1**, (see also *Precision Pendulum Clocks: France, Germany, America, and Recent Advancements*)
Riefler Company, 163
Ritchie, David, 96
Robin, Robert, 119 (see also *Precision Pendulum Clocks: France, Germany, America, and Recent Advancements*)
Robinson of Armagh, 54, **4-3**
Rudd, R. J., 133, 156
Satori, 104, 149
Schlessor, 145
Schuler, 145, 146, **7-9A-D**
Shelton, 181 (see also *English Precision Pendulum Clocks*)
Shepherd, 47, 148, 149, 152, **3-17A,B,C, 7-12A, 7-12B, 7-13, 7-14, 7-15, 7-16A, 7-16B**
Shortt, 54, 155, 158, 161, 165, **7-19, 7-20A,B,C** (see also *Precision Pendulum Clocks: France, Germany, America, and Recent Advancements*)
Steuart, Alexander, 153, **7-17A,B**
Strasser, Ludwig, 137 (see also *Precision Pendulum Clocks: France, Germany, America, and Recent Advancements*)
Strasser & Rohde, 137 (see also *Precision Pendulum Clocks: France, Germany, America, and Recent Advancements*)
Swartz, Franciscus, **6-1**
Synchronome Company, 156, 160
Thiout, Antoine, 119, 120
Thiout, L'Aine, 27, **3-9A,B,C**
Thompson, Sir William, 133
Thwaites & Reid, **5-32**
Tompion, Thomas, 10, 16, 17, 18, 21, 22, 23, 26, 27, 114, 168 - 180, **2-8, 3-6, 3-8A,B, 9-1, 9-2A,B,C, 9-3A,B,C, 9-4A,B,C, 9-5A,B,C, 9-7A,B, 9-8A-D**
Tompion & Graham, 169-189
Topping, John, **3-4**
Towneley, Richard, 16, 18, 20, 21, 22, 114
Troughton, Edward, 75, 93
Vulliamy, Benjamin, 116, **6-5B** (see also *English Precision Pendulum Clocks*)
Walsh, Henry, **6-6B** (see also *English Precision Pendulum Clocks*)
Ward, 88
Waugh of Dublin, 103
Webster, William, 188
White, James 133
Williamson of Ulverston, 93
Williamson, Joseph, 27, 37, **3-13A,B,C, 3-14A,B**
Winnock, Joshua, **6-2B**
Wright, Thomas, 187

ISBN: 0-7643-1636-2